"TO BE YOURSELF
IN A WORLD THAT IS
CONSTANTLY TRYING TO
MAKE YOU SOMETHING
ELSE IS THE GREATEST
ACCOMPLISHMENT."

RALPH WALDO EMERSON

ISBN
eBook version: ISBN: 978-1-9495-4200-4
Print version (Hardback): ISBN: 978-1-9495-4203-5
Print version (Paperback): ISBN: 978-1-9495-4202-8
Audio version: ISBN: 978-1-9495-4201-1

VERSION
Publisher: Justin Jones & Scott Waddell
eBook version: 1.0 | Published: September 24, 2018
Print version: 1.0 | Published: September 24, 2018

SALES
Got Ideas? How to Turn Your Ideas into Products People Want to Use can be purchased for educational, business, or sales promotional use. For more information, contact Justin Jones at justin@headflap.com or Scott Waddell at scott@waddell.build. Justin's website is Headflap.com. Scott's website is Waddell.build.

CONTENTS

JUSTIN JONES

Justin is a blue-collar redneck masquerading as a white-collar professional. He might be Director of Experience Strategy for a digital-product design agency in Salt Lake City, but at the end of the day, he heads back to his house in a rural outpost to relax by building bike trails on his property with his backhoe, a 1972 International that still fires up anytime he needs it. (The thing looks decrepit, but Justin used it to trench the power lines to his house.) You're far more likely to catch him in steel-toed boots and Carhartts than designer shoes and skinny jeans. He couldn't even begin to pull off a man bun.

JUSTIN IS NOT A CONFORMER.

He places high value on changing the way things "have always been done" and makes it a point to constantly question both rules and authority. He deeply believes that there is a better way to do just about anything and has an inability to sit with the status quo.

But despite his rogue nature, Justin is an excellence-seeker, a perfectionist who has worked in digital since AOL was the go-to ISP. He's passionate about digital — always has been — and has a knack for simplifying what others tend to overcomplicate. He thrives in chaotic conditions, needs tough problems to solve, and while he is considered a nightmare to work with by some, clients and colleagues who want to get products out the door always seem to call him back.

His work ethic comes from his upbringing in his father's heavy metal fabrication shop, where hard work was a number one value and even the boss's son had to prove himself, cleaning industrial shop toilets and welding back together broken dumpsters. It's here he learned to question everything, finding ways to save the company money by streamlining inefficient processes or building a better jig. There was a lot of sailor language on the shop floor. But the values Justin learned there have served him well throughout his life and career,

TEACHING HIM TO MOVE BEYOND THE STATUS QUO INTO THE REALM OF REAL VISION AND INNOVATIVE IDEAS.

SCOTT WADDELL

Scott's mother was often frustrated with him as a kid. He had a habit of taking things apart to see how they worked, deconstructing in order to reconstruct. Toys, old bikes, radios — he'd disassemble them and use a hacksaw, a drill, and bolts to put their parts back together in different ways to make them unique and better. He turned a television into a fish tank and a PEZ dispenser into a toothbrush, and combined a mousetrap and a spoon to make a catapult. He once took apart a brand-new push scooter the very day he got it, hacking off the back so he could extend it two feet and make an extra-long scooter capable of holding two kids at once.

A BUILDER FROM BIRTH,

Scott's heroes have always been imaginative makers of stuff: MacGyver, Doc from Back to the Future, James Bond's Q, Data from Goonies. He wanted to be just like one of his fictional heroes, a maker of amazing things.

Like many kids who are smart and creative, but different, he fell through the cracks at school. He didn't fit in socially, the classes bored him, his grades suffered. But in middle school, Scott discovered code. That changed everything. He found his passion, using his time in school to hone the craft of coding.

From there, the military taught Scott "adult skills." While coding is a valuable talent, so is getting to work on time.

The military taught him what it means to be accountable, how to handle conflict management, how to work on and with a team, and, most importantly to the life of a digital builder, how to embrace failure. Today, Scott considers success and failure to be essentially the same thing — just two expressions of learning.

Scott still gets bored easily. In civilian life, he looks for adventure in unexpected places. If someone tells him something is impossible, he considers that an intriguing challenge. He's committed to staying in the spirit of an explorer,

ALWAYS CARVING OUT HIS OWN TRAIL.

IMPLEMENTATION, YOUR SECRET WEAPON

So, you have an idea for a product. Now what?

We hear it all the time: "I have this great idea!" And we always respond the same way: "What have you done about it?" If the response is "Well, nothing" or "I'm waiting for a partner to come along," we know the person isn't serious.

To bring a digital product to life, you do need partners, but it's a chicken-versus-egg thing. No one wants to take the risk of working with someone who will turn out to be dead weight. Proving that you're capable of launching the development of a product is a big part of convincing people to work with you.

"ACTION EXPRESSES PRIORITIES."

GANDHI

We've coached many, many clients on the development of websites, apps, and advertising campaigns over the years, and we've found some of the most talented individuals and kindred spirits through collaborative implementation. Putting a project in motion is hard work, and the rewards aren't always clear at first. But looking back, it becomes obvious how new ideas, people, and products are illuminated through the journey of implementation.

If you have a good idea, start working on it now. That's how potential partners, advisors, employees, and investors will know you're serious.

> ## "IF YOU WANT TO CONQUER FEAR, DON'T SIT HOME AND THINK ABOUT IT. GO OUT AND GET BUSY."
> ### DALE CARNEGIE

Don't wait for a team to help you build. People are much more willing to throw a shoulder in and help once you've proven that you know how to get started. In fact, you might say that *implementation is your secret weapon*. It gives you a competitive edge. By implementing, you show you are not just serious, but capable.

Nobody wants to fail. It takes a leap of faith to create digital products, but the rewards are greater than the risks. Those rewards include the journey of learning as well as the potential that you'll end up with a successful product. We believe the things you will regret most in life are the things you wanted to do but didn't because you were afraid to take a chance.

Implementation is the process that turns ideas into actions, ideally in accordance with a plan. In this book, we'll walk you through the steps of implementation all the way to the finished product.

LET'S DIVE IN.

HOW TO READ THIS BOOK

If you're looking for a detailed textbook outlining every theory and practice of digital product design, this book isn't for you. Instead, this book is a synopsis of best practices and a collection of insights born both from our own experience and from industry truths.

Our approach to product design, which comes from our years of practical experience helping visionary product-makers get their ideas off the ground, as well as building our own digital products, IS ONE THAT REVOLVES EXCLUSIVELY AROUND FIGURING OUT WHAT USERS WANT, AND DOING THAT.

This nearly always results from an iterative process that isn't entirely linear — and so neither is this book.

While the components of the product-making process, and the various types of documents and tests that will help you with this process, are presented here in logical order, they are not necessarily sequential. That is, you might choose to skip around a bit in your own process, and there might be steps in this book that are not relevant to your particular product or company.

There are several ways you can read this book. You can absorb it cover to cover, in linear fashion, for a holistic, orderly overview of how to launch a product. If that's your game, you might consider using it in conjunction with our workbook, *Got Ideas? Work Them Out Here*, which takes you through this process step by step with specific instructions and the space to work out your plan.

Or you can use the table of contents to pick and choose subjects that you need more information about in the moment. We've kept the chapters short and concise, and even as they build on each other, they also stand alone.

Along the way, we've inserted threads of philosophy and words from the wise (some of the legendary digital product visionaries like Steve Jobs and Robert Noyce).

When it comes to the mindset behind developing revolutionary products, we are grateful to these greats. But we've also included plenty of anecdotes from our own personal experience working with digital product creators.

SO READ ON IN WHATEVER WAY BEST SERVES YOU!

"IF YOU LOVE WHAT YOU DO AND ARE WILLING TO DO WHAT IT TAKES, IT'S WITHIN YOUR REACH. AND IT'LL BE WORTH EVERY MINUTE YOU SPEND ALONE AT NIGHT, THINKING AND THINKING ABOUT WHAT IT IS YOU WANT TO DESIGN OR BUILD. IT'LL BE WORTH IT, I PROMISE." — STEVE WOZNIAK [1]

WE BELIEVE IN PEOPLE WITH IDEAS THEY BELIEVE IN.

It's a journey to bring an idea to life. It takes an immense amount of time, commitment, and willingness to sacrifice. And like anything hard, the rewards can be sweet and lasting. As Thomas Edison said, "Genius is 1 percent inspiration and 99% perspiration."

THE PRIDE AND SATISFACTION that follows the successful manifestation of an idea can only be experienced after you bring your product to life. When you hear people say, "It's my baby" about a project they really care about, there's a divine truth to that.

WE GIVE BIRTH TO THINGS; THEN THEY TAKE ON A LIFE OF THEIR OWN AND INFLUENCE OTHER PEOPLE'S LIVES.

---

1 Johnnie L. Roberts, *The Big Book of Business Quotations: Over 1,400 of the Smartest Things Ever Said about Making Money* [Delaware: Skyhorse Publishing, 2016].

When the very first iPhone model was released in 2007, it was groundbreaking, but there was a notable lack of video support. At the time, we were living on the bleeding edge of technology and working extensively in digital photo-to-video automation and encoding. When Apple's devkit went live in 2008, we saw an opportunity to infuse a video app feature, and we received one of the early days Apple developer invites. Aspirational even then, we set out to revolutionize the iPhone with video capability.

We had a lot of ideas about creating a web service to encode videos and make it easy to upload, manage, and publish them to the budding social networks. Our focus at the time was YouTube (still one of the most powerful social platforms to this day), and as we dove into video encoding and streaming, it quickly became an all-encompassing problem. But by mid-2009, Apple itself had gotten iOS on the full path toward robust video support. We were crushed, because in many ways, we thought we were solving the problem more quickly and better than Apple was. We lost a lot of time and money, but we were still left with some invaluable knowledge that has served our design thinking on every product ever since.

> ## "ONE WHO MAKES NO MISTAKES MAKES NOTHING AT ALL."
> **GIACOMO CASANOVA**

By the way, while we were spending all those late nights working on the iPhone video project, trying to solve a difficult problem, similar developers around the world were also trying to capitalize on Apple's devkit. In fact, one such developer, Joel Comm, managed to release an app called iFart Mobile that quickly shot to number one for app downloads and, for a time, pulled in nearly $10,000 a day. This was shocking to Apple, which had already nixed another novelty app called Pull My Finger because it seemed, well, dumb.

Unlike iFart Mobile, we lost a bunch of money on our efforts. But this lesson shaped the way we have viewed customer experience ever since and helped us figure out which products to build... and which ones not to build.

We've been building apps since that very first iPhone launch and have been involved in hundreds and hundreds of digital projects over the past decade. We've grown along with the industry and have held titles like Client Services Manager, Product Manager, Solutions Architect, Digital Strategist, Experience Strategist, Marketing Technology Director, and Vice President of Technology. We've had the immense privilege of working with hundreds of clients over the years on a breadth of digital projects like websites, apps, and ad campaigns. Living and working in the heart of Utah's booming and growing Silicon Slopes technology hub, we've seen many companies spring up over the past decade — some thrive, while most flail. And from this experience, we've gained deep insight into what makes a product truly excellent.

The first thing that makes something succeed is simply being willing to make it. We wrote this book for those on the precipice of creating a product who need a big shove over the edge and some solid advice on how to proceed. What we've found in our years of working with product designers is that the difference between those who dream and talk, and those who actually build things, boils down to a willingness to take that very first baby step.

We know, because we talked about writing this book for years. Writing a book was unfamiliar and unknown territory to us — not in our wheelhouse, so to speak — and taking that first step was not easy. Once we finally took it, we realized, holy smokes, this is going to be doable!

Most people, though, don't ever turn their ideas into actual products simply because they never bother to take the first step. Maybe they feel lazy about embarking on a huge journey. Maybe they are afraid to fail. Fear can be paralyzing. It's so easy to succumb to the voices in your head alerting you to all of the things that could go wrong, or why you can't afford it, or why your boss won't go for it, or that you just don't know where to start.

FEAR CAN PARALYZE US AND KEEP US FROM ACTING. **THE FIGHT-OR-FLIGHT MECHANISM KICKS IN AND WE STALL OUT, STUCK.**

Building products can seem like a daunting task if you think about everything that has to happen. But if you take it one step at a time, and surround yourself with the right like-minded people, it's not as massive of an undertaking as you might think. Not to say that you won't crash and burn. You might! But you have to be willing to fail, over and over again. This very thing is what separates the builders, creators, and doers from the mere dreamers and talkers.

"DON'T WORRY ABOUT FAILURE; YOU ONLY HAVE TO BE RIGHT ONCE."
DREW HOUSTON, DROPBOX FOUNDER[2]

Our hope is that this book will be the catalyst for you to take that first step. The rewards that come from creating and building things will help you become the best version of yourself and improve your outlook, career prospects, and value to your personal and professional communities. As you build momentum and continue taking baby steps, acting on ideas becomes a mindset, part of your core beliefs and values. It starts to shape who you are and how you see the world. We are passionate about user-centric products, and we hope that this book can offer a small nudge and hopefully demystify the magic that comes from creating products.

2   *"101 Best Inspirational Quotes for Entrepreneurs,"* Business Insider, accessed June 30, 2017, http://www.businessinsider.com/101-best-inspirational-quotes-for-entrepreneurs-2013-9

P.S. WE HAVE TO SAY THIS UPFRONT:

———

BUILDING THINGS CAN BE ALL-CONSUMING.

DON'T LET YOUR MISSION RULE YOUR LIFE.

BUILDING A GENIUS PRODUCT IS NEVER GOING TO BE MORE IMPORTANT THAN YOUR OWN HEALTH, YOUR FAMILY, AND SOME SEMBLANCE OF A BALANCED LIFE.

YOU MIGHT PUT IN SOME ALL-NIGHTERS AND SOME REALLY LONG HOURS, YES, BUT AT THE END OF THE DAY, GO TO BED KNOWING YOU SPENT TIME WITH YOUR FRIENDS AND FAMILY, TOO.

IN OTHER WORDS, PACE YOURSELF.

———

You can't build a house on fear

- Justin Jones -

Embracing failure is a methodology that extends far beyond the building of digital products. It's a mindset that leads to unconventional solutions in every area of life.

When my wife and I were finishing up our graduate degrees, we had a dream of building a modern house on a remote piece of land that would be our sanctuary from the city. We both loved our time spent at the University of Utah and wanted to stay close, but we also wanted a place in the wild to return to every night — a house with big windows and no blinds, where we could hear coyotes and owls in our backyard.

The Salt Lake City area is growing quickly, and properties like this sell for millions — which we didn't have. We decided to find a creative solution. Instead of scanning the local real estate listings, we started proactively contacting farmers and landowners in remote locations of the county. We drove all over, searched county records, and followed every unconventional lead. Our search ended on a barren 40-acre property bordering Bureau of Land Management (BLM) land. This parcel had been landlocked by BLM acreage for almost a century, making it difficult for the previous owners to access the property. No new buyer wanted to take a chance on it, so the price kept going down.

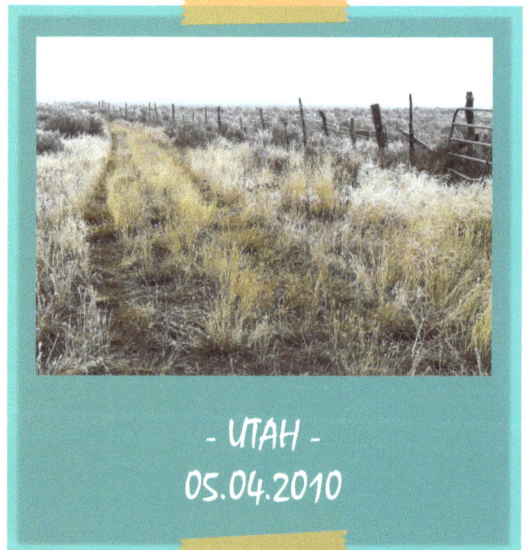

- UTAH -
05.04.2010

We decided to take a big risk on going head-to-head with the BLM to get the right of way. After several years of perpetual in-person visits, letters, emails, and relentlessness on every front, we received a perpetual easement across the BLM ground. This was only the beginning. Two more years of fighting the small town council (at the first meeting, one member proclaimed, "We are going to do everything in our power to prevent you from building!"), 600 feet of new road, a state highway right of way, a 455-foot-deep well, new power lines, several broken backhoe buckets, a year of 50+ hour all-nighter weekends, the pitching in of many rockstar friends, and a lot of luck, and our modern desert utopia was realized.

The risk of failure was high, but we knew the journey would be worth every minute. This same potential is what drives me every day I work on digital products.

1.0 THE FOUNDATION OF EVERY GREAT PRODUCT

This is a book about how to make things — specifically, digital products like apps, websites, APIs, and microservices. A digital product is a software-enabled product that solves a human problem. It might help you get paper clips delivered to your doorstep, catch a ride to the airport, or find out exactly how many steps you took today — all using a device like a smartwatch, laptop, tablet, or mobile phone. It creates value for a segment of users and also to the person or company that created it — and that value is not always monetary.

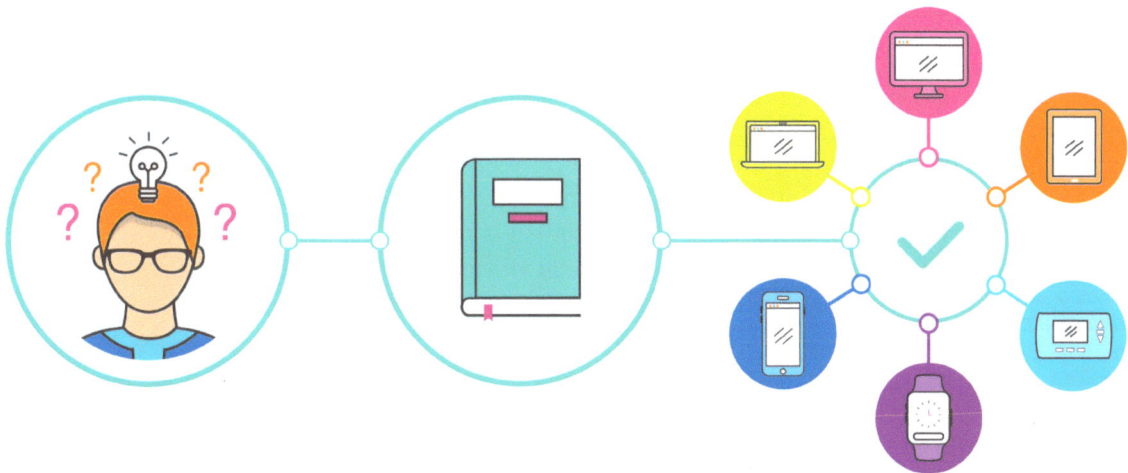

All you have to do is look at your smartphone (and if you live in the U.S., you probably have one on you right now — 77% of citizens do)[3] and you'll be fully immersed in the world of digital products. All of your favorite social media apps are right there: Facebook, Instagram, Snapchat, Twitter, LinkedIn. You also have at least one mapping app, and maybe you have Amazon Prime or another shopping app (or twenty) installed. You can check the weather, the local air quality, what time the sun sets today, and exactly what altitude you're at. You can read a book, watch a video, get the latest news. You can even shoot a film on your phone that looks like grainy old '70s footage.

APP DEVELOPERS HAVE THOUGHT OF EVERYTHING, RIGHT?
WRONG.

3 Aaron Smith, "Record shares of Americans now own smartphones, have home broadband," Pew Research Center, January 12, 2017, http://www.pewresearch.org/fact-tank/2017/01/12/evolution-of-technology

Developers submit over 1,000 new apps to Apple's App Store every day.[4] Amazon Web Services has created an entire ecosystem of cloud-based services and products. Not every digital product will be successful; most won't. But the velocity of product creation and the boundless wealth of ideas certainly means that there's still room for imaginative and bold product makers to thrive.

> ## "A UNIQUE KICK-ASS USER EXPERIENCE, ALIGNED WITH THE RIGHT BUSINESS MODEL, CAN DEFINE A 'DISRUPTIVE' PRODUCT."
> ## JAIME LEVY[5]

Every digital product consists of multiple digital touchpoints: specific tasks the user can engage in on that product. When you make a purchase from your Amazon shopping cart, that's a touchpoint, no matter whether you do so from a computer web interface or the Amazon app on your phone or tablet. Another type of touchpoint might be a search box embedded in an app or website. User experience is the culmination of touchpoints, along with the environment and perspective the user brings to the experience. A user's state of mind has a drastic impact on his or her experience.

> ## "FOR US, MONETIZATION ALWAYS CAME AFTER USER EXPERIENCE, AND THAT HELPED US MONETIZE LATER ON. YOU CAN'T BUY LOVE. YOU HAVE TO CREATE AN EXPERIENCE THAT PEOPLE LOVE AND NEED." — ADI TATARKO[6]

Think of a time when you've been driving and relying on Siri to help you out of a navigation pickle, only to have her plead ignorance to your plight. Or a time you've been trying to quickly log on to your online banking to get a figure while on the phone with a lender, only to have a pop-up block you and insist you fill out a "short survey" before you can access your account information. Frustrating. The emotional frame of a user is intimately connected to his or her experience with your product at various touchpoints.

---

4 Jerin Mathew, "Apple App Store growing by over 1,000 apps per day," *International Business Times*, June 6, 2015, http://www.ibtimes.co.uk/apple-app-store-growing-by-over-1000-apps-per-day-1504801
5 "Jaime Levy // What the Hell is UX Strategy? // UX Week 16," *Vimeo*, accessed June 30, 2017, https://vimeo.com/178863087
6 "Adi Tatarko," *Quotabelle*, accessed June 30, 2017, http://www.quotabelle.com/author/adi-tatarko

This book, boiled down to its essentials, is about user experience, or UX. User experience is at the heart of every successful project. It's why Apple is the first company to be valued at over $1 trillion, nearly four decades after it was founded. It's how Airbnb convinced people that renting to and from a complete stranger was a solid idea. It's how Uber made hitchhiking lucrative and relatively safe. And it's how Google innovated cloud collaboration with Google Drive.

UX strategy, at its core, is the intersection of UX design and business strategy. In other words, making it absolutely work for users, and also work absolutely well from a business perspective. If you can do both of those things, you have a viable product.

IN THIS BOOK, WE'LL GO DEEP INTO HOW TO ALIGN YOUR BUSINESS VISION WITH EXCELLENT UX AND USE LEAN PROCESSES TO GET A PRODUCT TO MARKET QUICKLY.

User Experience is all that matters

- Scott Waddell -

These days, I spend most of my time getting digital products off the ground. But once, I worked for a large wireless company as a Senior Wireless Engineer, charged with optimizing a customer-facing wireless network. We all know that what a wireless company says about its coverage matters not; it's whether we can make a call when we need to that we care about. The company knew this, too.

When a wireless company designs a cellular footprint in a given market, it starts with analyzing corridors of traffic — basically, where it expects customers to be. Then the company looks at the topology of the landscape, and attempts to place antennas in the right spots, at the right heights, to propagate signal strength into those areas. Once the equipment is in place, an army of testing vehicles drives around testing signal strength.

I started to notice that customers were complaining about their service in areas that, theoretically, should have had strong signal strength. Our optimization methodology hadn't worked well enough. We had to do better. We had to optimize from the customer's point of view.

To remedy this, I built an internal testing product revolving around a gadget with four different kinds of cell phones attached to it that could control a system, and we installed these gadgets in our army of testing vehicles. With these in place, we could understand how actual customer phones were interacting with the signal strength of the wireless network. Now, we could tell if a call dropped because of signal strength, if it was a wireless-to-wireless issue, or if it was a wireless-to-landline issue.

This was a big win for the wireless company, and it also transformed the way I began to look at product problems. From then on, I vowed to always look at every problem from the customer's point of view.

This wireless company has since been through several mergers and acquisitions since then, but the testing product I created is alive and well, still helping optimize the wireless network.

"IF I HAD ASKED PEOPLE WHAT THEY WANTED, THEY WOULD HAVE SAID FASTER HORSES." — HENRY FORD

At the outset, you must have a vision. Other things are important too, like understanding your market. You might have the greatest idea in the world, but if there's no market for it, there's no point. You must understand your potential users and their needs at a visceral level. You have to understand the demographics of your market so you know how to speak to potential customers. You must think about solving problems that your users aren't aware can be solved. And, of course, you must have a product idea that's in alignment with your own goals.

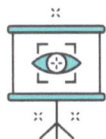    

But vision is what will keep your product on track throughout the entire life-cycle of your project. Without vision, you have nothing, and it will be almost impossible to maintain the immense amount of focus required to complete a digital product.

PRODUCT VISION IS SO MUCH MORE THAN JUST A GOOD IDEA. IT'S YOUR BIG PLAN, YOUR BLUE-SKY VIEW, THE ANSWER TO "WHAT COULD THIS THING GROW INTO?"

Product vision means having an intuitive understanding of what problem your product will solve for people and who those people are — whether or not they even know they have a need. It also means understanding how your product will grow and scale over time.

When Drew Houston first conceived the idea for Dropbox back in the early 2000s, he had a vision of a world where people wouldn't have to carry USB drives or email themselves attachments in order to synchronize and transport files. By this point, we had all gotten used to sticking a thumb drive into our USB port, manually moving files over, ejecting the drive, popping it in our pocket, and handing it to the guy at Kinko's or a client in a meeting. This actually seemed like a pretty modern solution, compared to floppy discs and CDs.

Houston imagined more for us. There were other such services in the space already, but most of them had terrible user experiences, so people weren't adopting them. Many of them didn't work across platforms, for instance. Houston saw an opportunity to radically improve the user experience of cloud-based file sharing. But many of the venture capital firms he approached scoffed at his idea, telling him that the storage market was too crowded. Like many product visionaries before him, Houston had an idea for a product that people didn't even know they needed, but that, once experienced, would change their lives.

We were early users of Dropbox. We met the founders (as well as their competitors) at TechCrunch, an annual event where technology startups launch their products on stage in front of potential investors. Along with a lot of other people in tech space, we sensed something missing with our file storage and sharing capabilities.

DROPBOX PROMISED TO SOLVE PROBLEMS WE WERE HAVING ON A DAILY BASIS:

- Trying to synchronize files across teams and disparate platforms (Windows and UNIX-based file systems)
- Trying to recover files accidentally deleted by design and product teams
- Trying to recover files when laptops were stolen or lost by employees
- Sorting through the mess of version control and inconsistent file-naming protocols (or, more accurately, lacks thereof)

We decided to try the product and were sold. One of our clients became Dropbox's largest business account for a while. In the years since, we've pitched and sold Dropbox to numerous clients, and we've done the math: Dropbox is currently receiving tens of thousands of dollars a year in direct revenue just from the companies we referred alone (and that doesn't count the companies those companies have probably referred). We've given numerous presentations and screencasts about how clients can better manage digital assets through Dropbox.

All of this business could have gone to another cloud file-sharing brand. But because we loved Dropbox's user experience the most, we've been loyal fans and vocal advocates ever since our early adoption. We don't work for Dropbox, but we are practically on their sales team (unpaid, of course). This is the power of good UX driven by unwavering vision.

PRODUCT VISION IS A LIVING CONCEPT. IT WILL CHANGE AS YOUR PRODUCT CHANGES. AND YOUR PRODUCT WILL CHANGE. BUT WITHOUT A PRODUCT VISION, YOUR PRODUCT IS DEAD FROM THE START.

"WE ARE STUBBORN ON VISION. WE ARE FLEXIBLE ON DETAILS."
JEFF BEZOS[7]

Did you know that Amazon.com was almost called Relentless.com? When Jeff Bezos originally envisioned his "online catalog" business, he registered the domain. The word relentless is a favorite of Bezos. He has been quoted as saying you need to be "incredibly relentless" with your product vision and believes that to truly accomplish anything important, you have to be "wrong a lot."[8]

Bezos and Amazon have been spectacularly wrong about a few things, like the Fire Phone, Amazon's entry into the smartphone arena, which debuted and flopped badly in 2014. But they've been right about a lot more, including that people, once indoctrinated into the idea of buying one thing online (in this case, books), would soon become used to buying just about anything online.

BEZOS LAUNCHED AMAZON WITH RELENTLESS VISION,
AND THAT'S WHAT YOU NEED, TOO.

---

7 Farhood Manjoo, "People Will Misunderstand You," *Slate*, August 21, 2011,
   http://www.slate.com/articles/technology/top_right/2011/08/people_will_misunderstand_you.html
8 Taylor Soper, "'Failure and innovation are inseparable twins': Amazon founder Jeff Bezos offers 7 leadership principles," *GeekWire*, October 28, 2016,
   http://www.geekwire.com/2016/amazon-founder-jeff-bezos-offers-6-leadership-principles-change-mind-lot-embrace-failure-ditch-powerpoints/

8 TIPS FOR PRODUCT VISION

Roman Pichler's list "8 Tips for Creating a Compelling Product Vision" provides a terrific way to think through your idea. Loosely paraphrased here:

1. **Describe the motivation behind the product.** What is your overarching goal?

2. **Look beyond the product.** Know the difference between your product and your vision.

3. **Distinguish between vision and product strategy.** Vision is a goal. Strategy is a road-map to get there.

4. **Choose an inspiring vision.** For instance, one that will solve a problem people have.

5. **Employ a shared vision.** Make sure everyone involved in building your product believes in the vision.

6. **Think big.** If a vision is broad enough, you have lots of room to change your mind within it.

7. **Keep your vision short and sweet.** But don't make your vision so ambiguous that it loses people.

8. **Use the vision to guide your decision.** Always, always, always filter your decisions through your vision.

YOU CAN READ MORE ABOUT THESE TIPS ON ROMANPICHLER.COM.

1.2 QUICK! CREATE YOUR VERY FIRST PROOF OF CONCEPT

> "I THINK IT'S VERY IMPORTANT TO HAVE A FEEDBACK LOOP, WHERE YOU'RE CONSTANTLY THINKING ABOUT WHAT YOU'VE DONE AND HOW YOU COULD BE DOING IT BETTER."
>
> ## ELON MUSK

You probably agree that product vision is essential to your project. But how exactly does this translate into action? Just like you can have a good product idea with no vision, you can also have incredible vision without a good product to anchor it, and the ability to execute on it.

SO, OBVIOUSLY, THE VERY FIRST THING TO CREATING A PRODUCT VISION
IS TO SCRUTINIZE YOUR PRODUCT IDEA.

Throughout this book we will talk frequently about the idea of iteration based on feedback. Getting your idea in front of people, taking in their feedback, and then making changes accordingly is how you will create a product people actually want to use. We will stress the concept of lean processes over and over and over again.

An example of a simple proof of concept

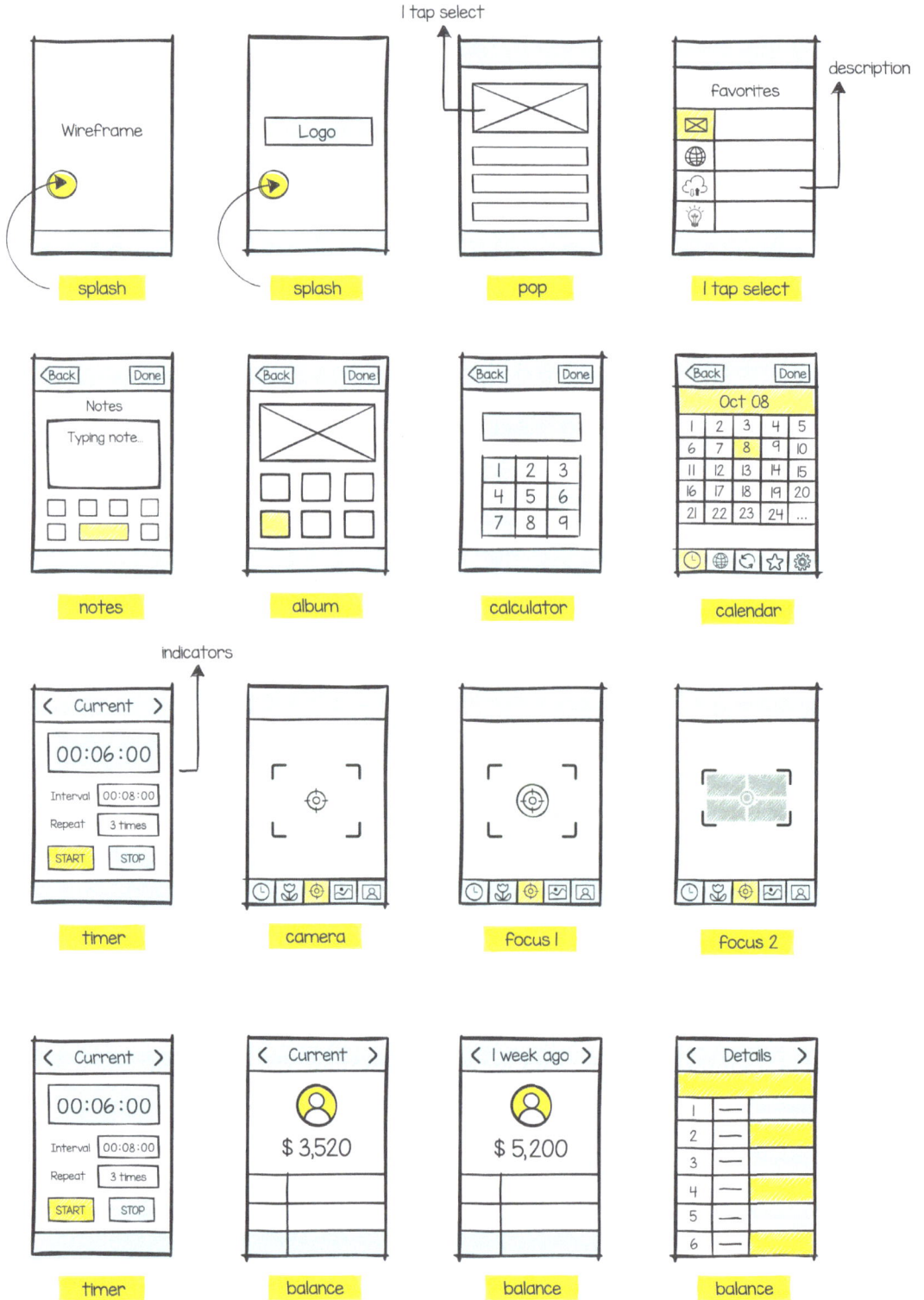

I tap select

Wireframe

splash

Logo

splash

pop

description

favorites

I tap select

Back Done

Notes

Typing note...

notes

Back Done

album

Back Done

1 2 3
4 5 6
7 8 9

calculator

Back Done

Oct 08

1	2	3	4	5
6	7	8	9	10
11	12	13	14	15
16	17	18	19	20
21	22	23	24	...

calendar

indicators

< Current >

00:06:00

Interval 00:08:00
Repeat 3 times
START STOP

timer

camera

focus 1

focus 2

< Current >

00:06:00

Interval 00:08:00
Repeat 3 times
START STOP

timer

< Current >

$ 3,520

balance

< 1 week ago >

$ 5,200

balance

< Details >

1	—	
2	—	
3	—	
4	—	
5	—	
6	—	

balance

And this starts with creating your very first proof of concept. The presentation doesn't matter; all that matters is the substance of the idea. Jot down some crude sketches of what your product might look like, screen by screen, on a napkin.

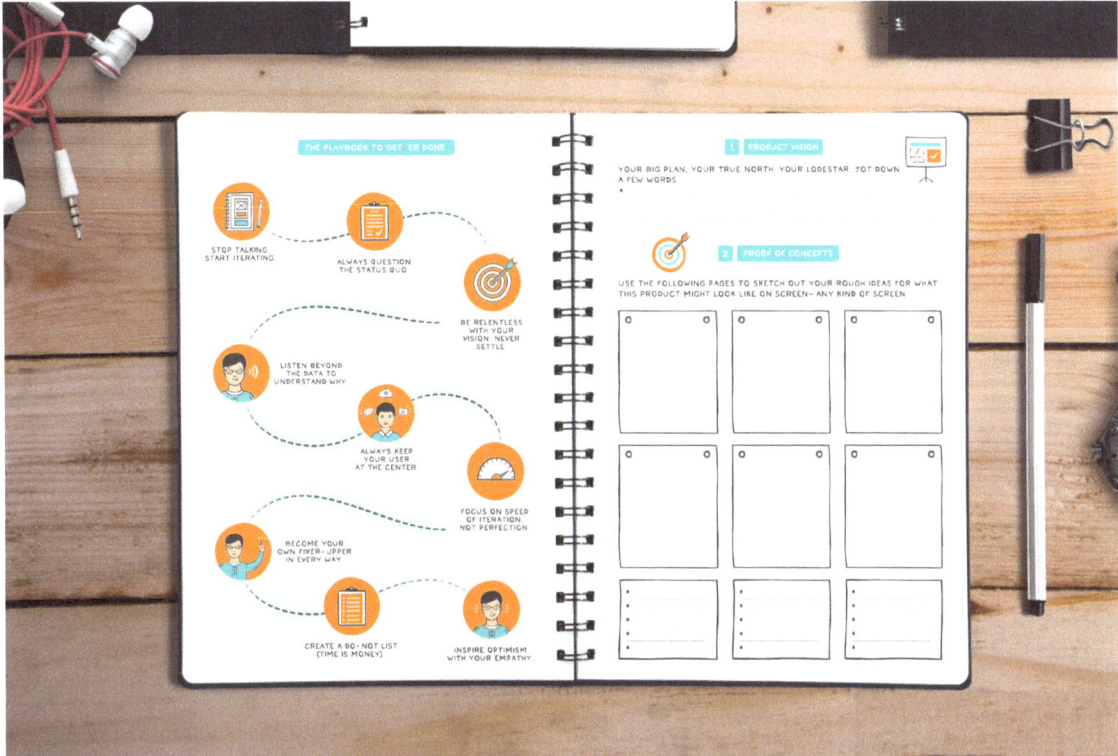

(Or use our product workbook, *Got Ideas? Work Them Out Here*, available on our site.) Show it to some people — maybe ten. Ask them for their feedback:

- Do you know what to do?
- Is it easy to use?
- How can we make this better?

Be willing to be humble about this. If you get all thumbs down, your idea is probably not very good. Close this book and come back to it when you have another one.

IF YOUR IDEA SPARKS INTEREST, BUT THAT INTEREST IS IFFY, GET SOME FEEDBACK. ITERATE YOUR IDEA. GET ANOTHER NAPKIN.

As you begin to riff on this very early stage proof of concept, always keep your vision front and center, developing features around it. If a particular feature seems like a good idea in theory, but doesn't jibe with your target market and its needs, you have two choices:

- Drop that idea.
- Change your product vision.

Getting back to Dropbox for a moment, as the founders were working through their vision, they posted a screencast on Hacker News to show how the product worked and to get feedback from users. The benefit of showing users a visual walk-through of the product — versus just trying to explain the idea in words — was that users got a sense of how to actually engage with the product.

FROM THIS, DROPBOX GOT A LOT OF VALUABLE FEEDBACK — BOTH POSITIVE AND NEGATIVE — THAT HELPED THEM SHAPE AND REFINE THE PRODUCT.

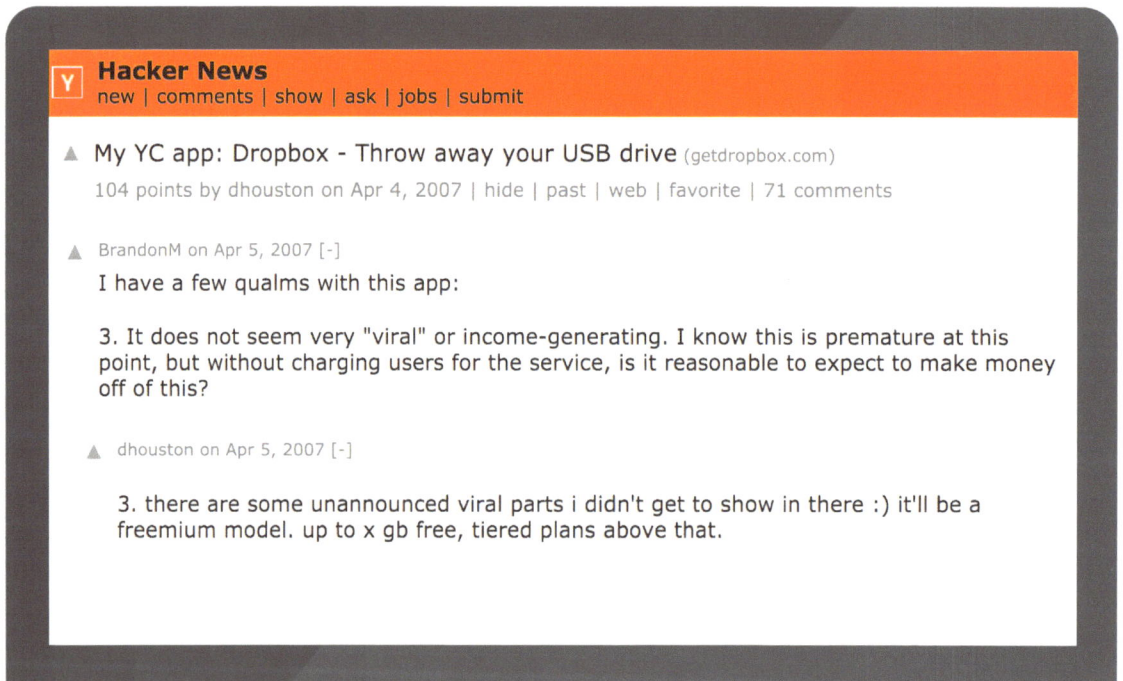

Y **Hacker News**
new | comments | show | ask | jobs | submit

▲ My YC app: Dropbox - Throw away your USB drive (getdropbox.com)
104 points by dhouston on Apr 4, 2007 | hide | past | web | favorite | 71 comments

▲ BrandonM on Apr 5, 2007 [-]
I have a few qualms with this app:

3. It does not seem very "viral" or income-generating. I know this is premature at this point, but without charging users for the service, is it reasonable to expect to make money off of this?

▲ dhouston on Apr 5, 2007 [-]

3. there are some unannounced viral parts i didn't get to show in there :) it'll be a freemium model. up to x gb free, tiered plans above that.

This Dropbox proof of concept was of the more sophisticated variety — more like an early stage prototype — but you get the idea.

A PROOF OF CONCEPT HELPS PEOPLE ENVISION YOUR PRODUCT, HOW IT WILL WORK, AND HOW IT WILL SOLVE A PROBLEM FOR THEM.

THIS EXERCISE IS ALSO A WAY TO GET VERY HUMBLE ABOUT ACCEPTING FEEDBACK, REALLY LISTENING TO IT, AND MAKING CHANGES TO YOUR PRODUCT AS A DIRECT RESULT.

Being able to do these things will serve you throughout the entire process of creating and marketing your product.

WITH YOUR VISION AND YOUR PROOF OF CONCEPT DONE, THE NEXT STEP IS TO DO SOME ACTUAL RESEARCH INTO WHETHER THERE'S A MARKET FOR YOUR PRODUCT.

THE BIGGER PICTURE

As you enter into the ambitious endeavor of envisioning your product, keep in mind the idea of product families. This might be your first digital product, but it probably won't be your last. As you work, always think about your product family potential and the product ecosystem you are creating.

As a caveat to this rule, we should mention that plenty of visionaries — Elon Musk, Richard Branson — don't really work this way. They have laser focus on one brilliant idea at a time, and if you look back on their lives' work, it's not necessarily on a theme, unless that theme is "really cool stuff."

But most business leaders in the digital world — yes, even Steve Jobs — are able to use their particular brand of focus to wrangle a fragmented and even outdated product lineup into a family of products that work together as a brand. And sometimes, those brands can even change the way we live.

1.3 MIND MAP YOUR PRODUCT BACK TO YOUR GOALS

How to get from vision to an actual product? The first tangible "deliverable" you should produce is a visual organization of your goals and ideas. And do it from your intended users' point of view. This early stage deliverable is a preemptive disaster check to ensure you're starting your project with a human-centric approach to creating features and solving problems. You need to make sure your product will solve a real problem for real humans — not an artificial problem you've created for an imaginary audience.

In our experience, we've found it's better to start with a solution to a narrow problem. Trying to solve too many problems at once usually results in feature bloat and bad UX.

By solving one problem, it then becomes much easier to expand and broaden your product and brand down the road. The flip side of that coin is creating a complex product with overly ambitious features, which inevitably subjects you to forced feature reduction — painful and a waste of time and money.

We recommend starting with a simple mind map. Mind mapping can be done on a whiteboard or using software like Lucidchart (there are plenty of other options as well). Your goal is to lay out your vision, write down everything you know at this early stage, and document the entire process. You want to get all of your knowledge out of your head and into a useful format you can share.

YOU CAN START WITH SOMETHING LIKE THIS, WHERE YOU LAY OUT YOUR PRODUCT VISION BROKEN DOWN INTO A FEW GOALS.

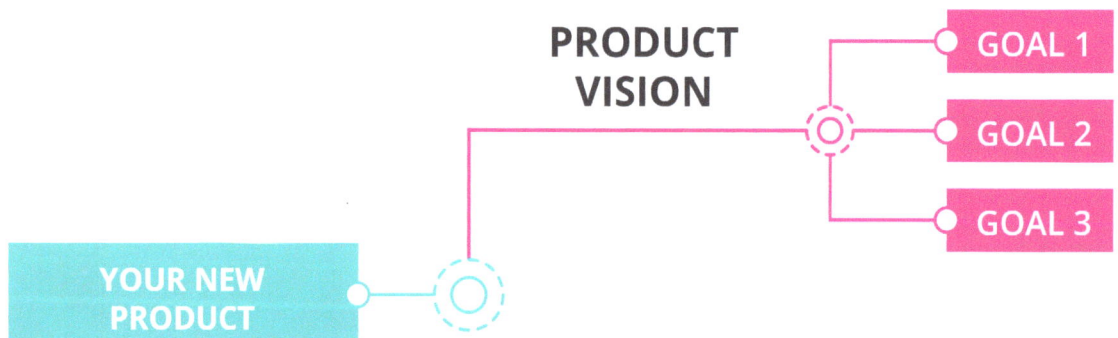

PRODUCT VISION

GOAL 1

GOAL 2

GOAL 3

YOUR NEW PRODUCT

If you are struggling with coming up with goals, sometimes it helps to define the tasks users will conduct with your product, then generate user goals from there:

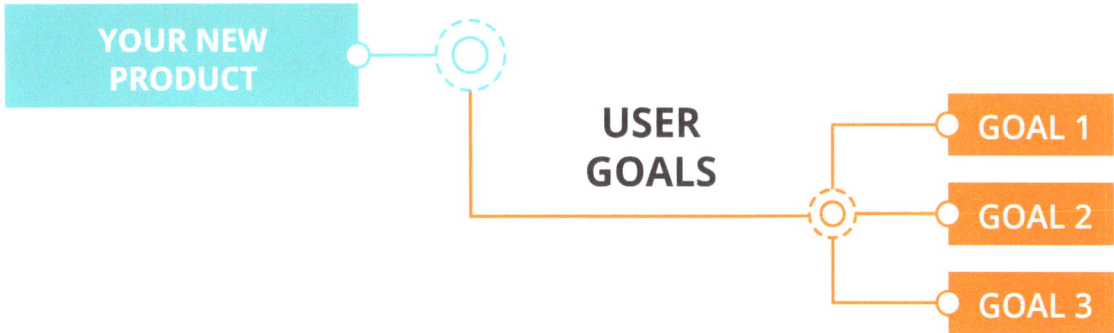

To illustrate the concept of a mind map, here is a contrived example of a very basic product idea: an app which will calculate percentages. Let's translate our vision for this into a designed workflow that will eventually turn into a prototype. Here is the start of a mind map with product vision goals and user goals contrasted:

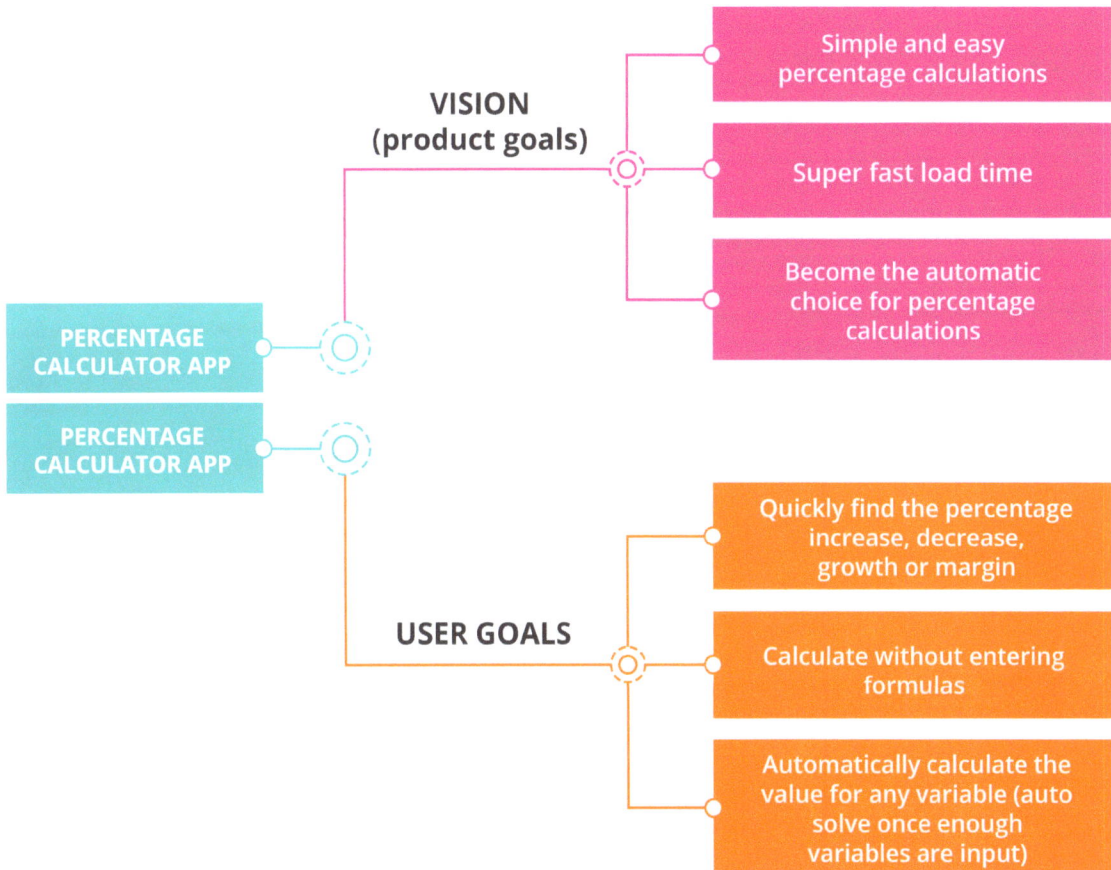

Notice that none of the goals include things like app usage, downloads, profits, or number of users. These things might be important, but they aren't goals; they're metrics, and you can track them after your product is released. They should not be driving your vision for user experience up front. This, of course, is one of the major pitfalls of many organizations. They start the process with the goal of making millions... or at least getting millions of users. They want to create the next big thing, and they dream about the potential return on investment (ROI). But these types of business drivers should not influence your initial prototype at this early stage. Right now, your goal should be to make an awesome product that people need and love. The rest will follow.

At this point, the best way to start is to brainstorm all the features that you've considered or thought about so far. Conduct a thorough data dump, which will likely get you a list of features much longer than this example we're using.

YOUR PRODUCT IDEA IS PROBABLY MORE COMPLEX THAN A SIMPLE PERCENTAGE CALCULATOR, AFTER ALL.

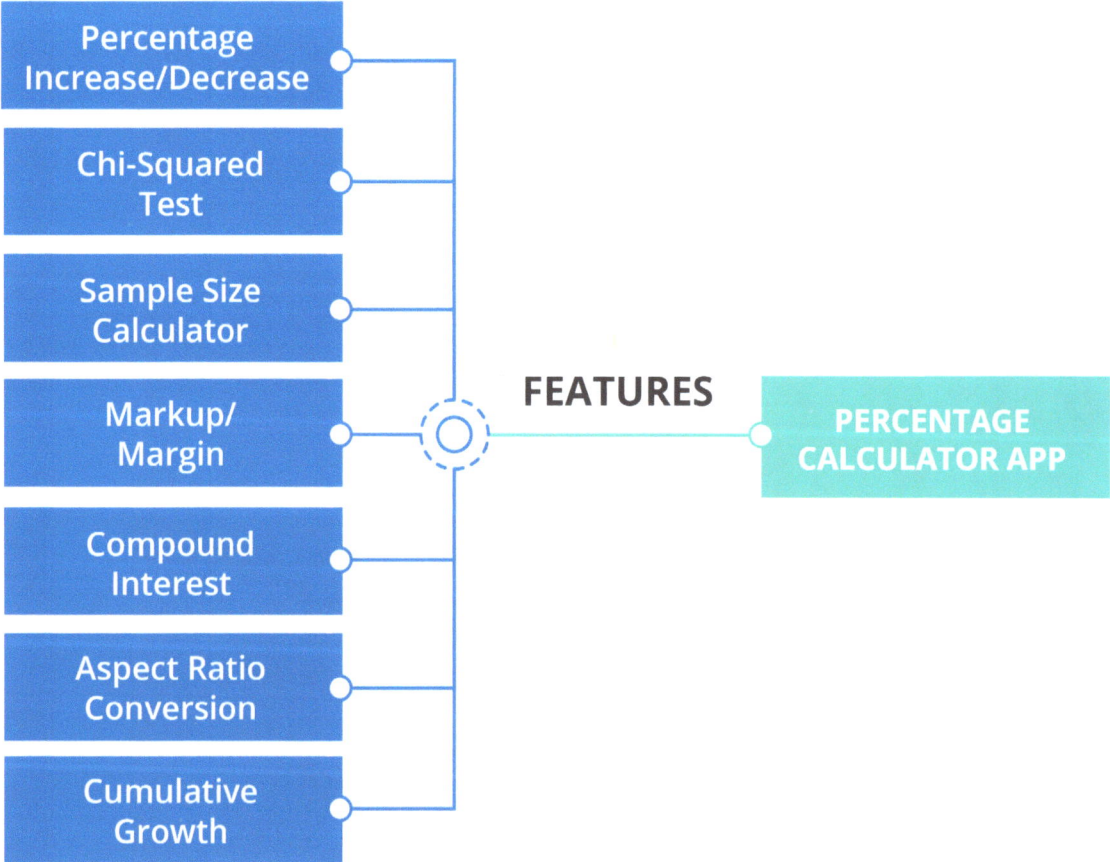

Percentage Increase/Decrease

Chi-Squared Test

Sample Size Calculator

Markup/ Margin

Compound Interest

Aspect Ratio Conversion

Cumulative Growth

FEATURES

PERCENTAGE CALCULATOR APP

Once you have that data dump down, go through an exercise of feature bashing. Make sure every feature you've come up with aligns back to one of your goals. If it doesn't, eliminate it. But don't stop there. The next step is to define which of the features left standing are actually essential right now. If a feature is not essential to your early prototype, put it in your "nice to have" folder for later.

Right now, your focus should be on critical features so you can nimbly iterate on your idea.

BY THE END OF THIS EXERCISE, YOU SHOULD HAVE A CLEAR IDEA OF HOW YOUR PRODUCT WILL LOOK AND WHICH KEY FEATURES SHOULD BE DEVELOPED FIRST.

YOU'RE GETTING CLOSER TO A FIRST PROTOTYPE.

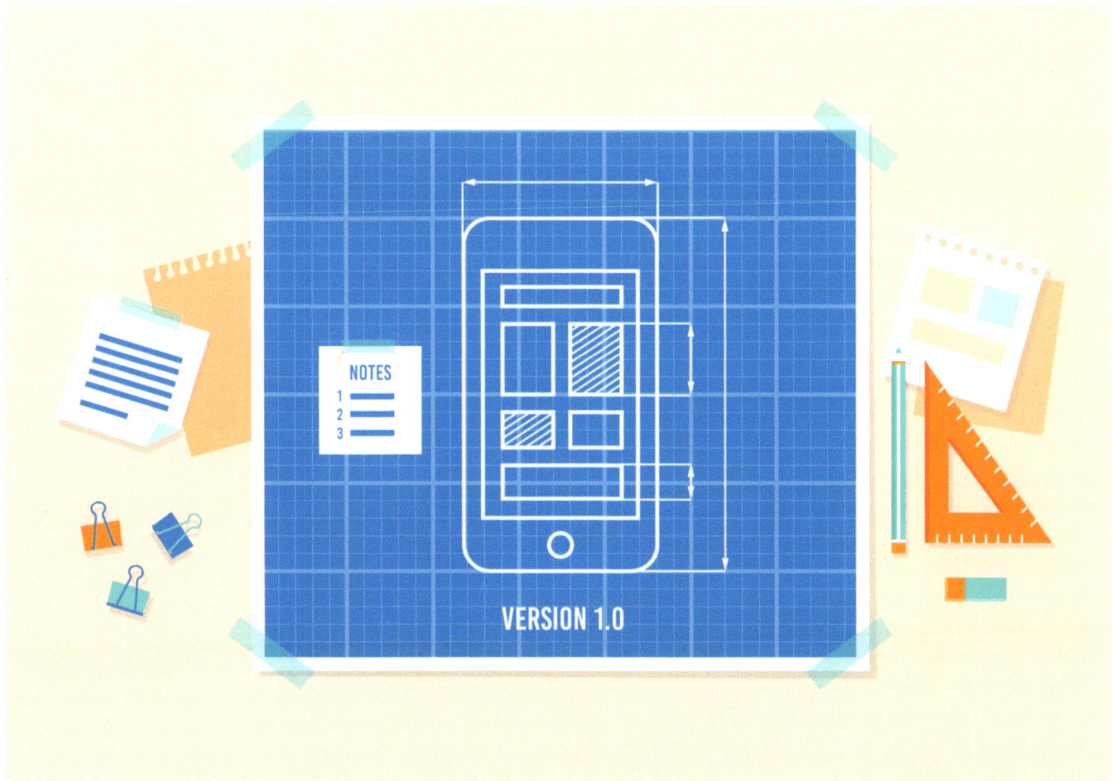

"OPTIMISM IS AN ESSENTIAL INGREDIENT OF INNOVATION. HOW ELSE CAN THE INDIVIDUAL WELCOME CHANGE OVER SECURITY, ADVENTURE OVER STAYING IN SAFE PLACES?"
ROBERT NOYCE[9]

With a mind map down on paper, you can slide easily into the next step: a workflow diagram of your product. Some designers call these "wireframes," and we've heard other names, too. We like to keep it simple. The focus of this exercise is to outline key activities, tasks, and workflows between your product's screens.

IF YOU'RE FAMILIAR WITH UNIFIED MODELING LANGUAGE (UML) DIAGRAMS, THIS IS A SIMILAR IDEA, BUT MUCH SIMPLER.

---

9 "Robert Noyce, Statesman of Silicon Valley," *Intel*, accessed June 30, 2017,
   https://www.intel.com/content/www/us/en/history/museum-robert-noyce.html

WHAT A WORKFLOW DIAGRAM MIGHT LOOK LIKE.

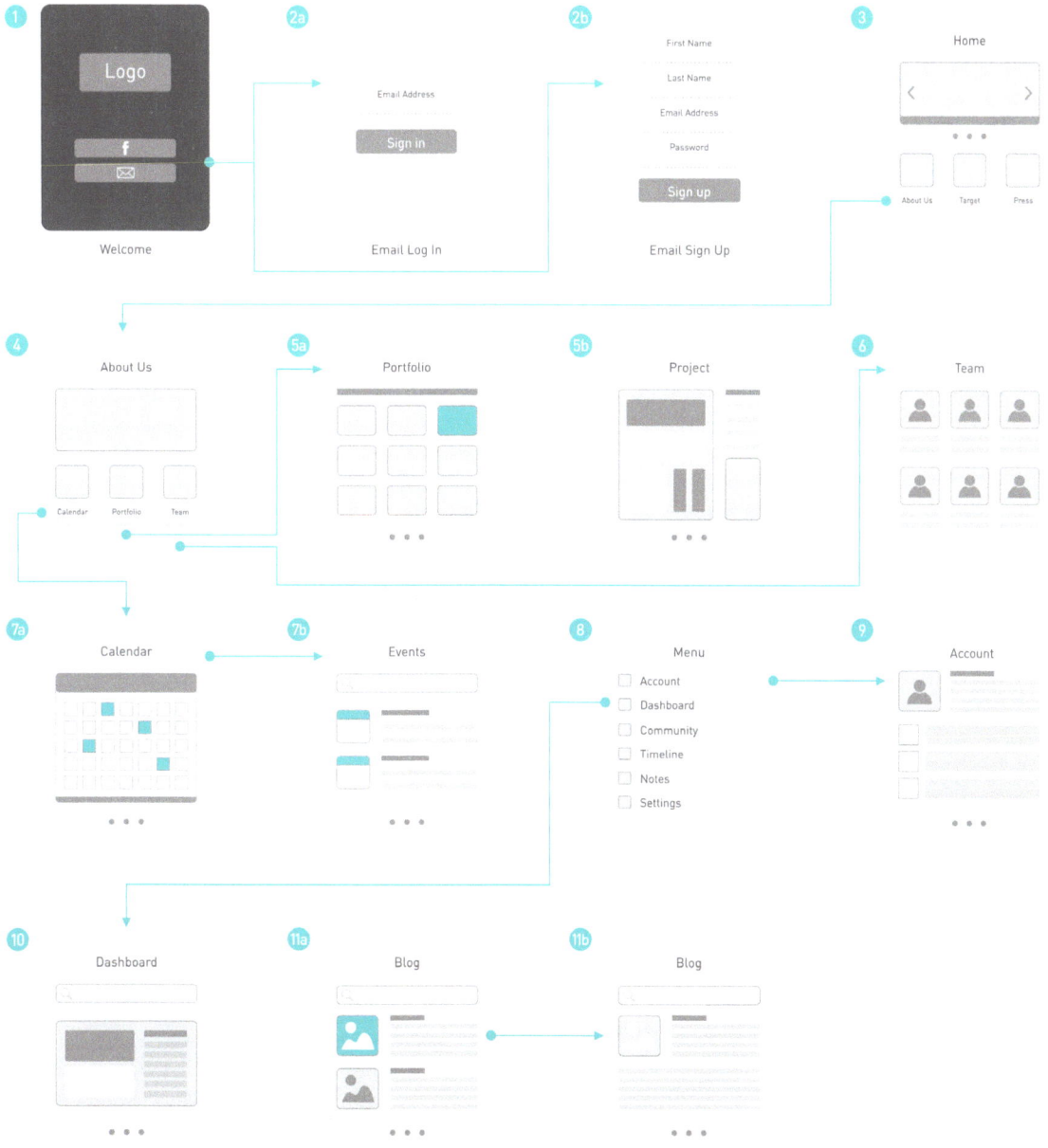

1 Logo

f

Welcome

2a Email Address

Sign in

Email Log In

2b First Name

Last Name

Email Address

Password

Sign up

Email Sign Up

3 Home

< >

About Us   Target   Press

4 About Us

Calendar   Portfolio   Team

5a Portfolio

5b Project

6 Team

7a Calendar

7b Events

8 Menu

☐ Account
☐ Dashboard
☐ Community
☐ Timeline
☐ Notes
☐ Settings

9 Account

10 Dashboard

11a Blog

11b Blog

The essence of a product workflow diagram is to outline what you need to design and how it will interconnect as the user navigates through the interface. This will help you connect all the pieces: screens, buttons, actions, etc.

Workflow can look visually overwhelming at first, so the key is to keep it simple at first and then iterate as you go. Have an eraser in hand as you work, so to speak.

If you're a pro who does this type of work every day for clients, you might be used to all types of modeling notations and tools. But for most entrepreneurs, this is just one step in a long journey, and not something they do every day. There are a lot of tools you can use. Choose whatever seems easiest and works best with your style. You could, for instance, use the same tool you used for mind mapping — specific mind-mapping software or a whiteboard.

Another good technique for diagramming product workflow is to use Post-it notes. These little slips of paper are easy to move around and align on a large board. Similarly, UXKits. com has several tiny wireframe templates and a deck of UX cards that are easy to lay out on a table when designing responsive website development. Here's an example:

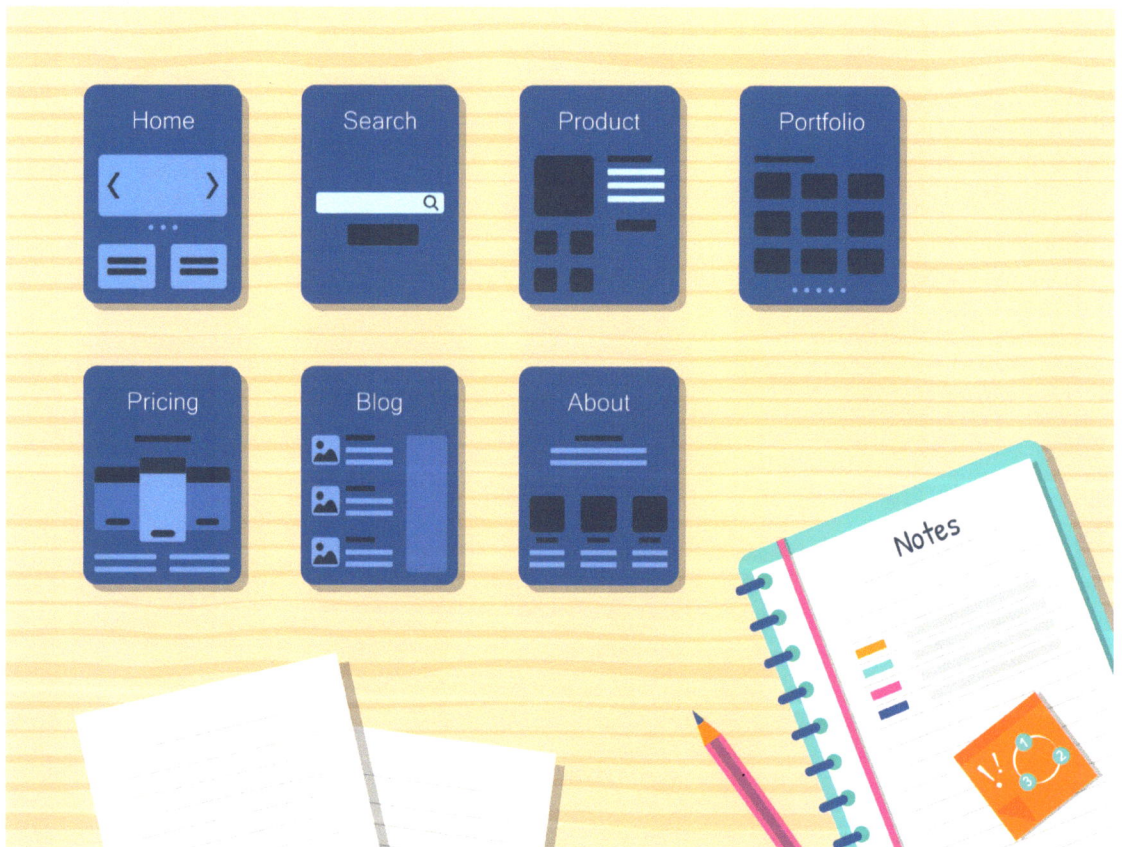

UX CARDS CAN HELP YOU DESIGN A RESPONSIVE WEBSITE.

Last but not least, two tools that we find ourselves using again and again are good old PowerPoint and Keynote. These presentation software tools are the current love language of most business and product people, as well as stakeholders and executives. They're surprisingly efficient for this kind of work.

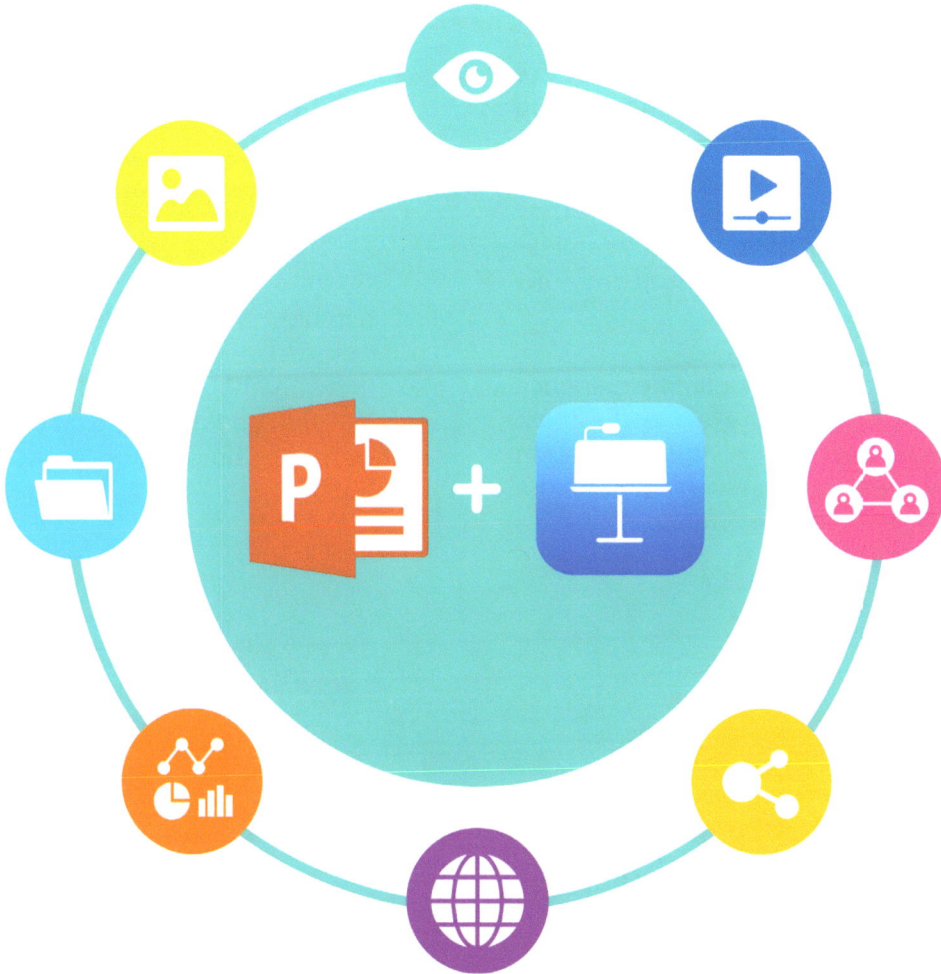

As you work through the workflow diagram exercise, one of your main goals should be to identify the core experience of your product for the user. For example, with a web app, the core experience is the page where the user tends to spend the most time. On Facebook, for example, the core experience is the News Feed. On a blogging platform like Medium, rather than the homepage, the core experience for the user is probably the article view, regardless of which article a person is reading.

MOVING INTO THE NEXT STEP — MAPPING THE USER JOURNEY — YOU'LL KEEP THIS CORE EXPERIENCE IN MIND.

1.5 MAP YOUR USERS' JOURNEY

Today's customer journey is complex, full of both digital and physical touchpoints. Customers research, review, and use products on various mobile and desktop devices throughout their day. Regardless of which device they are on, they expect instant answers, advice, and action. They want their experience to be seamless across devices, and as they stop and start their journey.

In order to create such a fluid experience for customers, and to execute the most successful purchasing and marketing paths, it's critical that you identify the touchpoints that are most likely to motivate customers.

TOUCHPOINTS ARE KEY INTERACTIONS YOUR CUSTOMER WILL HAVE WITH YOUR PRODUCT AT VARIOUS TIMES.

For instance, the customer might visit your website one day, favorite a tweet the next, download a white paper the third, and then sign up for a demo on the fourth — all before he buys your product.

"MOST ORGANIZATIONS ARE REASONABLY GOOD AT GATHERING DATA ON THEIR USERS.

BUT DATA OFTEN FAILS TO COMMUNICATE THE FRUSTRATIONS AND EXPERIENCES OF CUSTOMERS.

A STORY CAN DO THAT, AND ONE OF THE BEST STORYTELLING TOOLS IN BUSINESS IS THE CUSTOMER JOURNEY MAP."

PAUL BOAG ON SMASHING MAGAZINE[10]

---

10 Paul Boag, "All You Need to Know About Customer Journey Mapping," *Smashing Magazine*, January 15, 2015, https://www.smashingmagazine.com/2015/01/all-about-customer-journey-mapping/

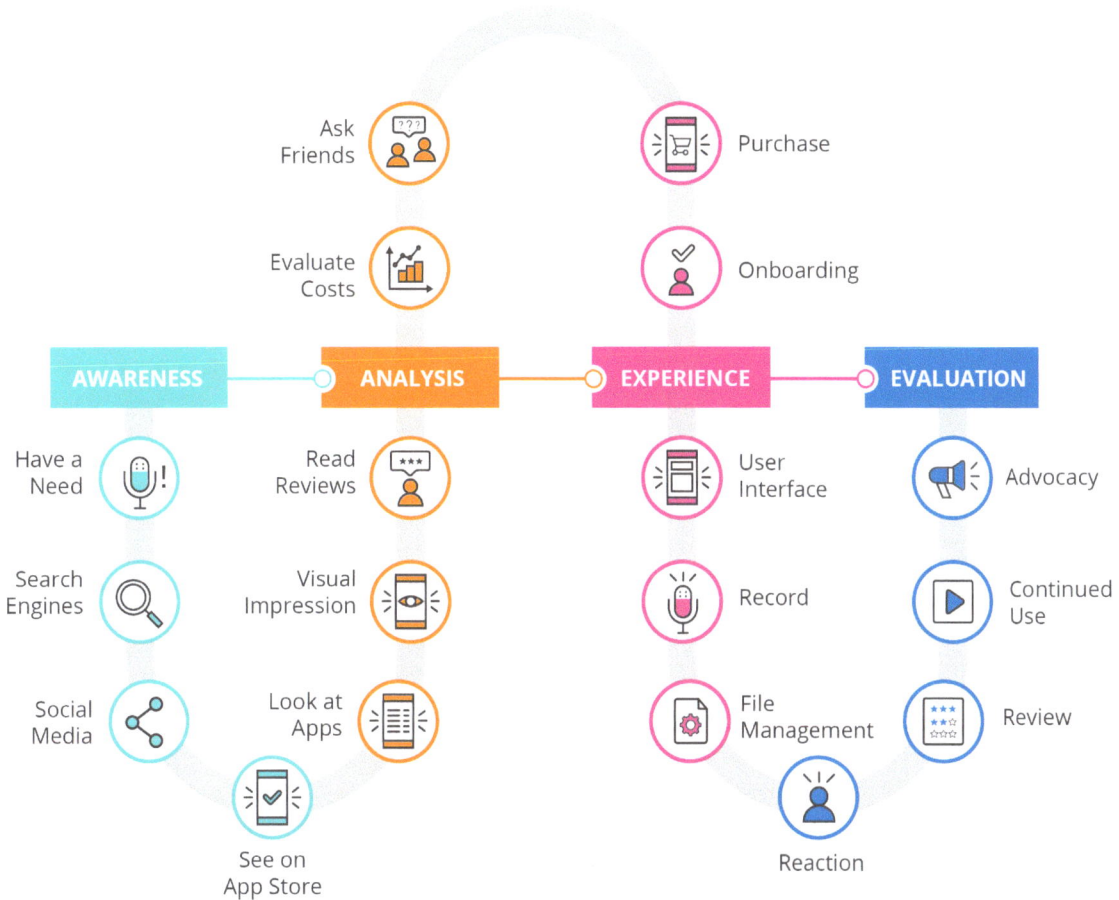

AWARENESS
- Ask Friends
- Evaluate Costs
- Have a Need
- Search Engines
- Social Media

ANALYSIS
- Read Reviews
- Visual Impression
- Look at Apps
- See on App Store

EXPERIENCE
- Purchase
- Onboarding
- User Interface
- Record
- File Management
- Reaction

EVALUATION
- Advocacy
- Continued Use
- Review

Defining these touchpoints, and the user's motivations and questions at each, is called mapping the user journey. The user journey map is a graphical tool for everyone involved with your project, from engineers to designers to copywriters. This mapping process helps identify gaps in your user experience, and it also increases your ability to generate awareness about your product, enlighten customers to its benefits, and motivate them to take action. When you're working toward landing investors, it's crucial that you understand this aspect of your product.

The customer journey map eventually becomes a key document in identifying problems with your product. Once your product is developed, it also becomes the foundation of your media plan.

MOST IMPORTANTLY, THE JOURNEY MAP BECOMES THE CATALYST TO CREATING A USER-CENTRIC PRODUCT.

It will be heavily used in determining your future product roadmap and ensuring you are staying true to solving actual customer needs and problems.

As you plan out your product roadmap, you should put specific emphasis on optimizing each touchpoint systematically. As you go out and talk to users, continue to refine the journey map until it's boiled down to the key elements that are important to your users and product. Once you have identified these points, it becomes the foundation of effective testing strategies and enables you to chunk up complex user flows into simple (and more actionable) tests.

Owners of products that are already on the market have a major advantage when it comes to mapping the journey of their users. They have existing data in the form of website and social analytics, anecdotal research gained from speaking with customers, and the ability to conduct tangible research with things like customer surveys. But when you're creating a brand new product from scratch, you have to use your imagination a little bit.

On your first pass at creating a user journey map, try and account for every touchpoint you can think of. As you move forward, refine and combine the touchpoints to ensure your product experience aligns with your customer journey. Rather than looking at the customer journey with a typical funnel model, we recommend looking at it as a linear path. Connect the digital and physical touchpoints and plot them in relation to where they would land in the funnel.

Throughout this book, you'll notice a theme of less is more, and the customer journey map is no exception. You should be able to create it quickly — within a few hours.

DON'T WORRY ABOUT MAKING IT PERFECT; IT WILL BE A DYNAMIC DOCUMENT, UPDATED AS YOUR PROTOTYPE EVOLVES.

GOOGLE'S DIGITAL TOUCHPOINT TOOL

"MOBILE HAS FOREVER CHANGED THE WAY WE LIVE...
IT'S FRACTURED THE CUSTOMER JOURNEY INTO
HUNDREDS OF REAL-TIME, INTENT-DRIVEN MICRO-MOMENTS.
EACH ONE IS A CRITICAL OPPORTUNITY FOR BRANDS
TO SHAPE OUR DECISIONS AND PREFERENCES." — GOOGLE[11]

If you're struggling with coming up with touchpoints, take on the mindset of a user and come up with a few statements that start with "I want to ___." Then, trace the moment at which the user's need is met with your product marketing. Google calls these types of interactions micro-moments. For brands, capitalizing on them means being able to quickly meet consumers' needs in the moment, wherever they occur.

Google has an online tool you can use as a starting point to map your customer journey. Select an industry, business size, and region, then see how different marketing channels affect online purchase decisions according to Google Analytics data. This is a pretty rudimentary tool, so you won't want to rely on it as your sole mapping device, but it can give you some quality initial insight.

GO TO: HTTPS://WWW.THINKWITHGOOGLE.COM/TOOLS/

---

11 "Micro-Moments," *Google*, accessed June 30, 2017,
https://www.thinkwithgoogle.com/micromoments/intro.html

1.6 NOW, TRANSLATE YOUR VISION INTO USER-CENTRIC DESIGN

"CREATIVE PEOPLE HAVE AN ABIDING CURIOSITY AND AN INSATIABLE DESIRE TO LEARN HOW AND WHY THINGS WORK.

THEY TAKE NOTHING FOR GRANTED. THEY ARE INTERESTED IN THINGS AROUND THEM AND TEND TO STOW AWAY BITS AND PIECES OF INFORMATION IN THEIR MINDS FOR THE FUTURE USE.

AND, THEY HAVE A GREAT ABILITY TO MOBILIZE THEIR THINKING AND EXPERIENCES FOR USE IN SOLVING A NEW PROBLEM."

BILL HEWLETT[12]

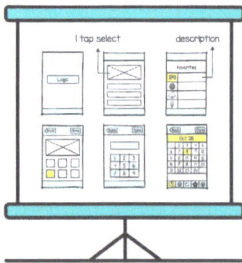

We've worked with a lot of designers, which is how we've come to realize that diagramming product workflow doesn't jibe with every type of brain. Designers are highly visual, and so are lots of other people, by the way. If you're one of those people, moving straight from mind mapping to design is the path of least resistance. Your goal is to do so in the shortest amount of time possible and to come out of this exercise with the most realistic, albeit faked, prototype you can pull off.

---

12 Michael Shawn Malone, *Bill & Dave: How Hewlett and Packard Built the World's Greatest Company* [Portfolio Hardcover, 2007]

When we say prototype at this early stage, of course, we are often referring to things like sketches on paper. We're not talking about a functional, working app. This level of design is rudimentary but vital to your future product. And that product, by the way, might be an app or a website or something else. This chapter applies to any type of digital product you're creating.

DON'T GET TOO CREATIVE

A mistake we often see at the early design stage is the impulse to do everything from scratch. You have a revolutionary idea — it's never been done before! — so it needs a brand new type of design, right? No.

A lot of really smart people have done a lot of really good work when it comes to UX design. There are incredible libraries of open-source stock images you can pull design elements from for now. Two we recommend are Unsplash and The Noun Project. Files from both can be pulled directly into your prototype.

» CHOOSE YOUR TOOLS WISELY «

There are a lot of tools you can use for this design exercise: Sketch and Adobe's Creative Cloud (which includes Photoshop, Illustrator, and particularly Adobe XD) are a few. Whatever you choose, our advice is to start with a grid, a component library, and a design system based on the front-end framework. Let's break down what this means.

GRID SYSTEM

A grid system will help you place your design elements in a way that's orderly and visually appealing. Some old-school designers balk at grid systems because they feel it limits their creativity. But we recommend using a grid because it helps when it comes time to translate your design into code — especially code that's responsive and works on multiple types of screens. It's a much more efficient process and ensures your prototype can be developed more easily and cheaply.

A twelve-column grid is the most common for websites and front-end frameworks, because it's divisible by 1, 2, 3, 4, and 6. This gives your design lots of options: you can combine six columns to use half your page, or use three columns for 25 percent. You can also nest within the grid. So twelve works nicely, but it's not the only way.

On the other end of the spectrum, your grid could be as simple as setting up guides in a document to ensure consistent sizing, spacing, margins, etc. The best designers we work with make extensive use of guides, patterns, and libraries.

MOBILE

LAPTOP

DESKTOP

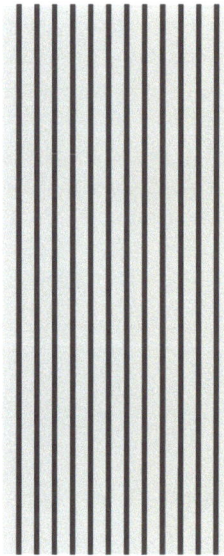

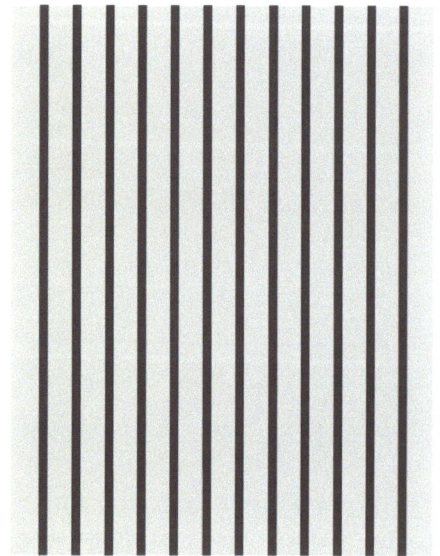

A grid system can help you place design elements to work across different types of displays.

GRID RESOURCES

For easy downloadable grids for any size or type of device, visit Behance.net. For a tutorial on how to use grids in your design, visit Foundation.zurb.com.

COMPONENT LIBRARY

Once you have a grid in place, you can layer on UI with a component library — essentially a group of building blocks that might include basic UI elements like glyphs/icons, alerts, menus, drop-downs, labels, and form fields. We typically recommend component libraries that are built on a front-end framework because when it comes time to enter development, these building blocks are already coded. There are also a ton of customized resources, both free and paid, based on these front-end frameworks, that you can use to further customize your design with pre-styled components. This will accelerate your dev time by a large margin.

One caveat with component libraries: the taxonomy isn't always clear on the web, and sometimes you might hear component libraries referred to as UI kits. But in our mind, they are two different things.

COLORS

NAVBAR

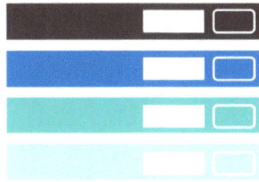

TYPOGRAPHY

Display 1
Display 2
Display 3
Display 4

IMAGES & FIGURES

CARDS

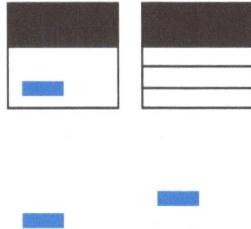

MEDIA OBJECTS

An example of a component library.

COMPONENT LIBRARIES

We recommend component libraries built on a front-end framework, like Bootstrap or Foundation. Bootstrap has a component library specifically for Experience Designer and one for Sketch.

DESIGN SYSTEM

For apps, we can't emphasize enough the importance of leveraging a design system up front for the platform you're building toward (which we'll talk about more in 1.7). This is a comprehensive design resource package that helps you create a fluid experience across platforms and devices, from iPhone to Apple TV to Apple Watch to Mac computer, for instance. They give you specific guidelines around user experience for things like motion, animation, and sizing. They often include a developer library or portal.

These systems usually have their own custom fonts built specifically for the platform in question: Roboto (Google), San Francisco (Apple), Salesforce Sans (Salesforce). They also have detailed UI elements, app controls, navigation controls, progress bars, toolbars, buttons, and other interface elements unique to a platform or device.

For Android, iOS, Salesforce, and Windows, these design systems have been in development for years, with continual optimizations, and they should be the jumping-off point for your app. They give you a foundational design to work from — a mix-and-match ability to pull items from the kit for each page or screen. Apple has its own Human Interface Guidelines (HIGs) that cover all of its platforms, including Apple Watch and Apple TV. Google has Material Design that covers motion, style, layout, components, patterns, communications, usability, and platforms.

Starting with this framework for design will be something your development team will greatly appreciate in the future, and it will create immediate efficiency for your design team in the near term, too. At this early stage of development, showing user flow is much more important than beautiful photo blending or custom vectors. That's why we usually lean toward a rudimentary tool like Experience Designer or Sketch, two of the best prototyping tools on the market. Both were built for simplicity, speed, and ease of mocking up an experience.

DEVELOPER LIBRARY

DEVICES

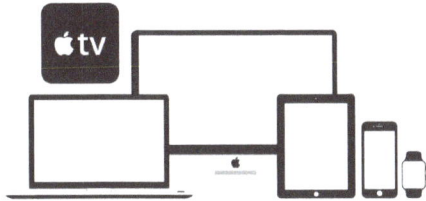

SERVICES

UX GUIDELINES

MOTION

ANIMATION

SIZING

ZOOMING

PLATFORM GUIDELINES

CAMERA

VIDEO

GPS

SOUND

There are a ton of resources you can turn to for a UI kit template. For Android, material.io has a good resource library, and for iOS, we recommend checking out Apple's HIG, which include downloadable templates. The latter includes Photoshop, Sketch, and Adobe XD templates and other materials that will help you quickly construct iOS apps.

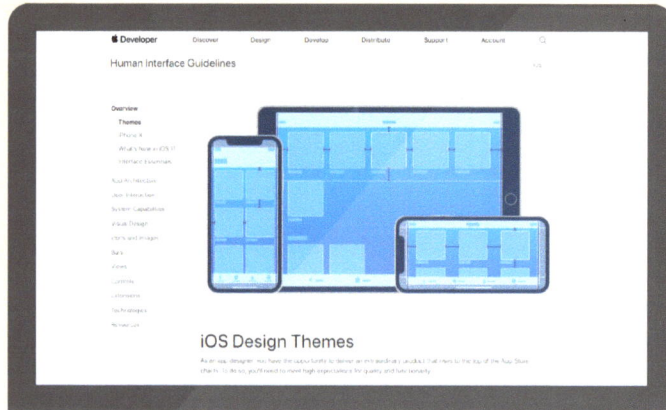

Apple's HIG

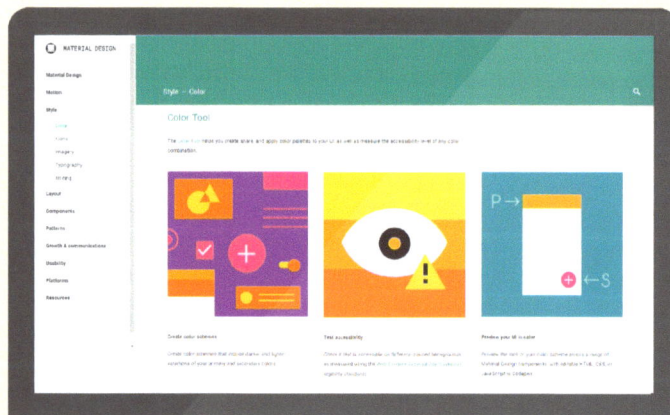

A resource library for Android app design on material.io

OTHER DESIGNER TOOLS YOU MIGHT CHOOSE

If your design team prefers a presentation product like PowerPoint, just make sure you're simulating the size of your ultimate screen view. In other words, if you're designing an app, the screens in your presentation should look like those of a mobile device. There are several ways to achieve this by inserting screenshots into the file. You can use the "responsive view" in your web browser to dial in on exact dimensions — a function most browsers have these days — and take screenshots that way (see box). Alternatively, you can use a device simulator or simply take screenshots off your mobile device and insert them into the presentation file.

Once you have these screenshots, it's easy to overlay shapes and UI elements to get the correct dimensions, spacing, text sizing, and design patterns. An entire mobile app can be prototyped in a few hours this way.

Design tools like Sketch and Experience Designer will produce a more polished-looking workflow of your product than presentation tools will. It's also far easier to create links — linking buttons or actions on a page to show where they will go. But if you're wanting to highlight animations, add sounds, create dynamic in-page interactions with your prototype, presentation tools have a distinct advantage.

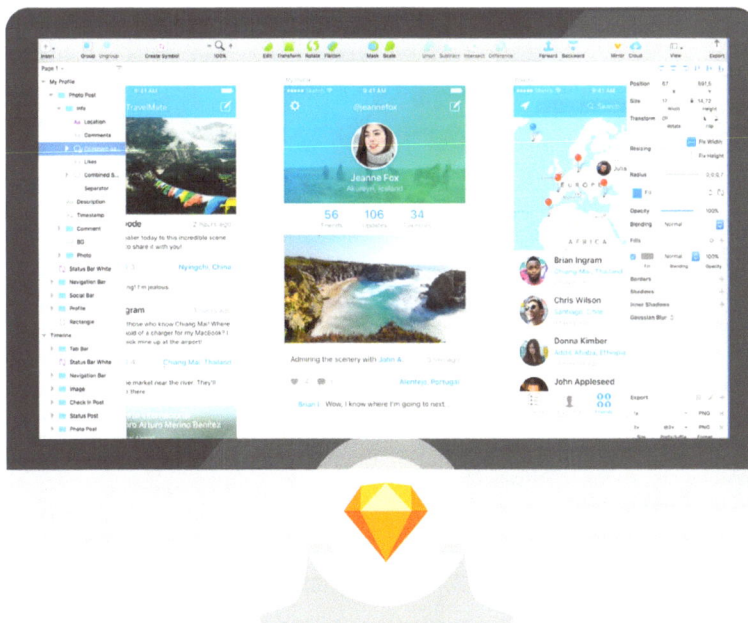

The Mac digital design tool Sketch, which makes it easy for anyone
to create plugins and easily market them

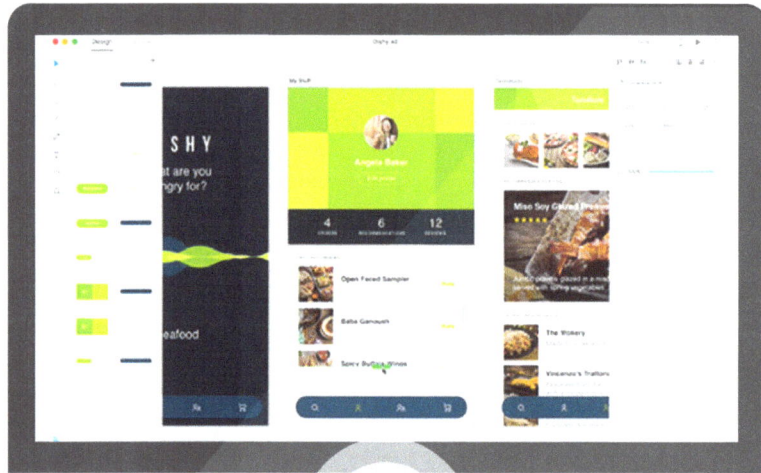

The more robust (and pricey) Adobe's Experience Design CC, which makes digital design and prototyping easy with an entire design toolkit that aims to simplify the workflow for UX designers

HOW TO FAKE MOBILE VIEW

One way you can achieve mobile screenshots for your design prototype is to change the view of your browser to display at mobile size, even from a desktop or laptop browser. In Google Chrome, for instance, you would hit the following key combination:

Command+Option+I (Mac) | F12 or Control+Shift+I (Windows / Linux)
To see this pop-up:

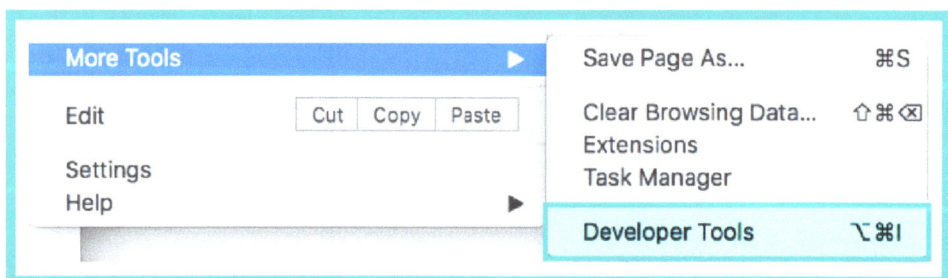

By choosing Developer Tools, you can select a device to preview, with or without the frame visible:

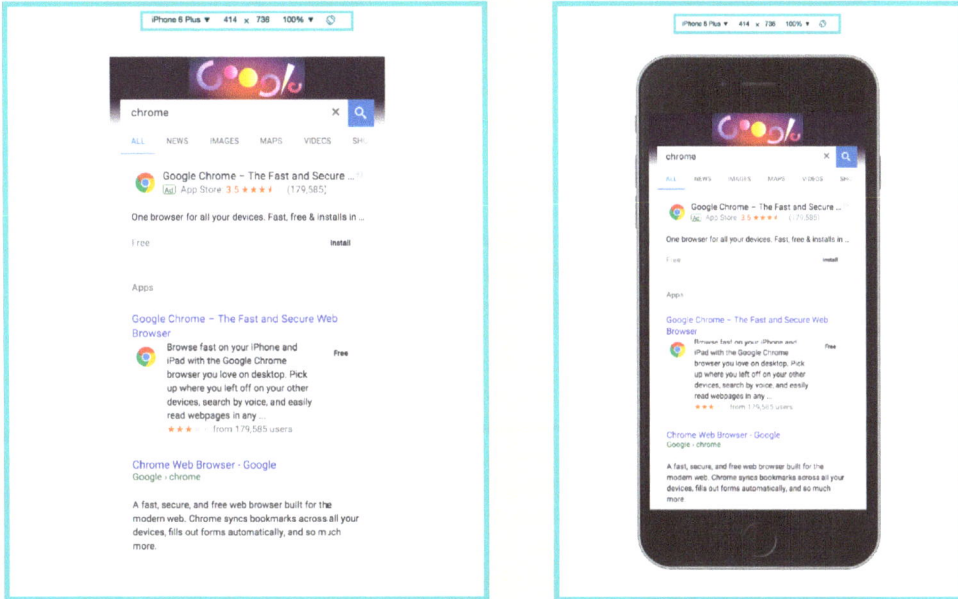

» PUT YOUR TOOLS — AND MINDS — TO WORK «

Always start with what you know. In this case, you know the essence of your product in terms of your stated vision and goals. Take the time to think through the primary experience of your product in depth. What are the elements, tasks, and screens that are essential?

Once you have an initial pass done, you should create a contrarian view as an alternative. For instance, if your mockup has lots of options and is text heavy, go in the opposite direction and make an iteration that reduces the text and options, and maybe introduces imagery or iconography instead. We recommend doing a minimum of three iterations — preferably you do five or more, always focusing and asking yourself the question "Is there a different, better, or new way of illustrating this experience?"

It's pretty typical for designers to have a preconceived notion of what experience they want, and forget that they are not the user. They'll go out and build a ton of screens based on their own personal biases. When designers are enlisted, the mindset can get even further away from the end user and delve into the realm of the beautiful app that really isn't all that useful.

It's always important to focus on the user, but in the beginning, it's dire, because the direction you take here will enormously influence the way your product ends up later on. If you realize later that you didn't consider the user view, it takes a lot more time to go back and alter workflows and designs. Often, a major change down the road will mean all your early effort spent on auxiliary elements was wasted. This makes designers cranky. They will become irritable, push back, and sometimes leave you backpedaling on a change you know is right because you want to preserve the relationship.

TO ELIMINATE ALL THIS HEARTACHE, FORCE YOUR TEAM TO ITERATE EARLY ON.

If you have more than one designer, having them each create a few iterations doubles your ideas and creative options and also forges a healthy spirit of competition.

» ALIGN YOUR DESIGN WITH YOUR VISION «

"THE WAY TO GET STARTED IS TO QUIT TALKING AND BEGIN DOING."
WALT DISNEY

Finally, bring your key people back together to conduct a design and vision review. Hopefully, your team can take constructive feedback and set their egos aside. This is a must-have skill for innovative product development. All of your team members' credentials, college degrees, personal pompousness, and attachment to time spent on the mockups must be shelved. You're in search of the best idea. The one that delivers on your vision wins.

CAREFULLY ANALYZE AND EVALUATE EACH OF YOUR DESIGN ITERATIONS.

DOES IT FEEL RIGHT?

DOES IT ALIGN WITH YOUR VISION AND GOALS?

DOES IT FULFILL THE USER'S GOALS?

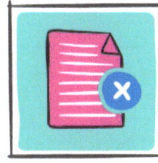

HAVE YOU ELIMINATED ALL FUNCTIONALITY THAT DOESN'T SUPPORT YOUR STATED GOALS?

DOES IT HAVE TOO MUCH/TOO LITTLE INFORMATION?

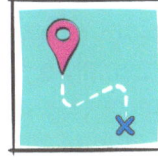

DOES IT GUIDE THE USER?

IS IT INTUITIVE?

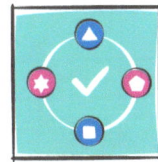

DOES IT USE ELEMENTS AND PATTERNS USERS ARE FAMILIAR WITH?

DOES IT REFLECT YOUR BEST THINKING?

IS THERE TOO MUCH TEXT?

ARE THE IMAGES SIZED RIGHT?

ARE THERE TOO MANY IMAGES?

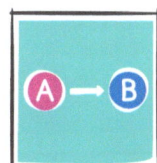

IS THERE A SIMPLER WAY TO DO IT?

These questions will inevitably lead to valuable discussions. Depending on the outcome of this initial review, you may decide to have the designers take another pass, or you may feel good about a direction. When there are two valid points of view, ideally, you foster an environment with smart compromise — although you shouldn't force someone to compromise for compromise sake.

At this stage, as the product owner, you should decide on one option to pursue. For most apps, the crucial step is getting the core experience right. Then, the other pieces fall into place much easier. Sometimes, though, you may have different segments of users or experiences, and in those cases we recommend repeating this process for those experiences. Once you have gone through this on the core experiences, then you can enable your design team to build out a more comprehensive workflow based on the established pattern.

In the last section of this book we talked about how a workflow diagram comes together. This is a similar process in some respects, but instead of a diagram on paper, the end result is a clickable, or at least showable, prototype. This is when the magic starts to happen. It might take a week or two to get there, but with a lean team, you might be able to pull it off much quicker — in a day or two. We usually try to keep team size pretty small at this stage of a product. By small, we mean four or less people. If you can manage to get a product this far with that few people, you're doing something right, and your stakeholders will be duly impressed. We'll talk a lot more about team size later in this book. Consider this a teaser.

WHAT IF MY PRODUCT IS SUPER COMPLEX?

We are always harping about keeping things super simple. But sometimes a product is complex by its very nature. The trick is to recognize when you're making a product overly complex and when the complexity of features actually benefits the product and its users.

For instance, say you're designing a peer-to-peer app for buyers and sellers of motorcycles. There are two obvious workflows: the one for buyers, and the one for sellers. Or perhaps you're creating a blogging platform. You have two types of customers: those creating blogs, and those reading blogs.

For any project that is going to result in isolated customer experiences like these, we recommend making smart splits. Break them into manageable parts and work on those parts one at a time.

As you're doing this, you have to keep your holistic vision in mind and make sure that each part ties back to that vision.

1.7 BOW DOWN TO THE PRINCIPLES OF HUMAN-CENTRIC DESIGN

"WHEN YOU INNOVATE YOU'VE GOT TO BE PREPARED FOR EVERYONE TELLING YOU YOU'RE NUTS."
LARRY ELLISON[13]

We're all surrounded by design, and often we don't even realize it because the most well-designed experiences are invisible.

IT'S ONLY WHEN A PRODUCT'S DESIGN IS BAD THAT WE TEND TO NOTICE IT.

It gets in our way, confuses us, or hinders our ability to do something.

User experience is at the core of product success. Good UX might seem like a basic concept, but creating intuitive and useful product experiences is immensely challenging, as any product manager will attest to. One of the primary challenges any new product is up against is that it's going to be compared to the most widely used and carefully refined digital products in the world — most of which have simple and intuitive interfaces.

---

13 "10 Classic CEO Quotes for Today's Economy," *CBS Money Watch*, last update September 28, 2009, http://www.cbsnews.com/news/10-classic-ceo-quotes-for-todays-economy/

Not every bad UX fails right away, either. At one time, MySpace was considered the number one internet property, ahead of even Google. MySpace had tons of resources, plenty of users, celebrity endorsements, and a huge lead in market adoption. Mighty MySpace blazed a path for social media to succeed. Suddenly, out of nowhere, the original social media site was squashed by a little upstart called Facebook.

WHY? FACEBOOK WAS JUST BETTER.

MySpace had lost sight of what it means to be a human-centric product. Facebook had a laser focus on what users really wanted.

And remember that Facebook started as a simple tool for college coeds to learn each other's names and faces. It took a few major pivots for the company to become the all-knowing, all-seeing data juggernaut it is today. Through it all, the common vision has been an interface people love.

As you enter into these early stages of design, get your mind wrapped firmly around the concept of how to design with a real user in mind. At this stage in your product design, it's time to do a design disaster check. Stop what you're doing, and measure your workflow or your design (depending on which route you went) against the following principles of design.

SIMPLICITY

"SIMPLICITY IS NOT THE ABSENCE OF CLUTTER, THAT'S A CONSEQUENCE OF SIMPLICITY.

SIMPLICITY IS SOMEHOW ESSENTIALLY DESCRIBING THE PURPOSE AND PLACE OF AN OBJECT AND PRODUCT.

THE ABSENCE OF CLUTTER IS JUST A CLUTTER-FREE PRODUCT. THAT'S NOT SIMPLE."

JONY IVE[14]

14 Shane Richmond, "Jonathan Ive interview: simplicity isn't simple," *The Telegraph*, May 23, 2012.

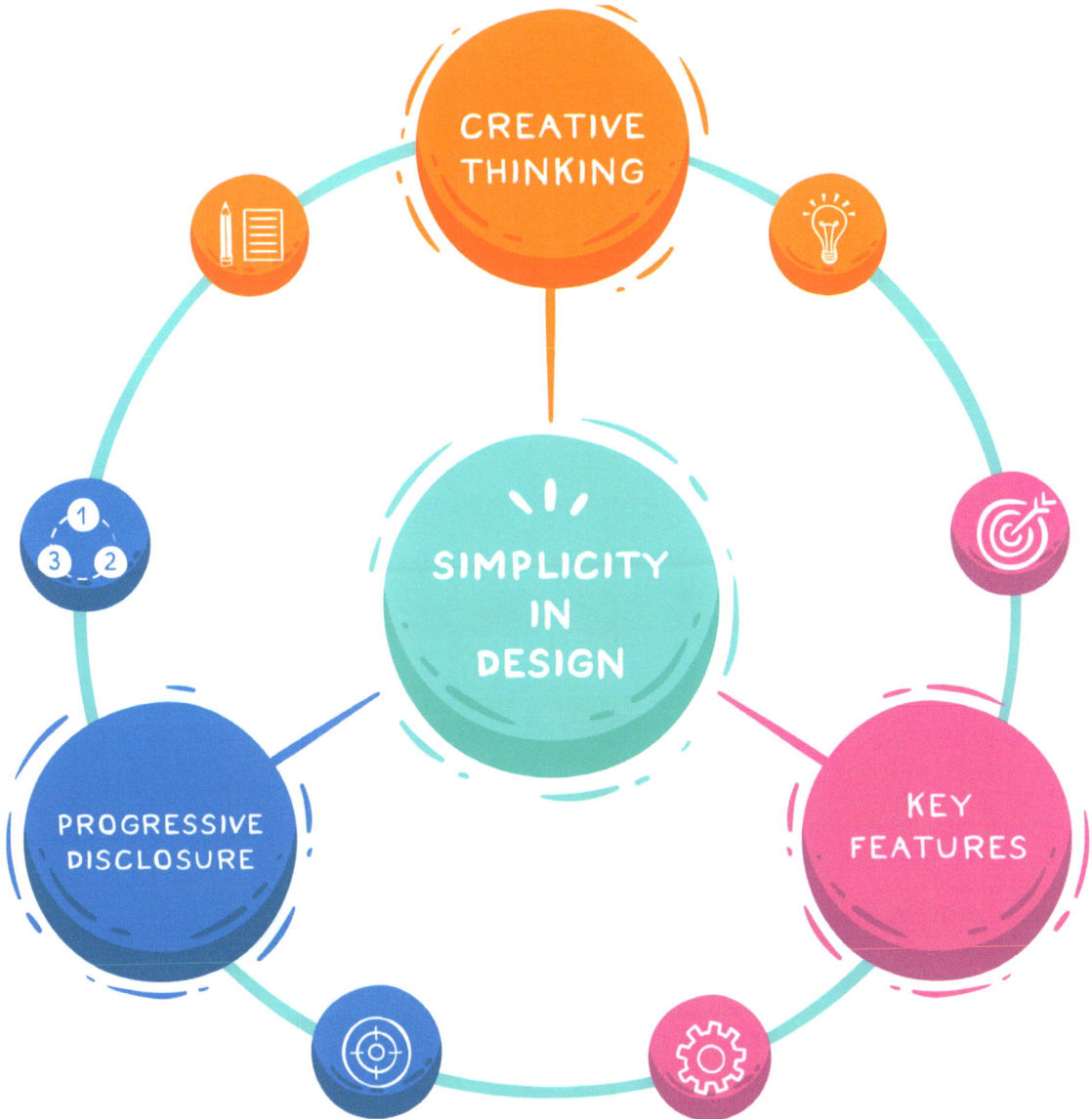

EMBRACING PROFOUND SIMPLICITY IS THE MOST IMPORTANT TRAIT OF HUMAN-CENTRIC, USER-FRIENDLY DESIGN.

Leonardo da Vinci said, "Simplicity is the ultimate sophistication." But simple can be more difficult than complex because it requires a deep understanding of the complex in order to simplify.

Simplicity is synonymous with clarity, visibility, reliability, and maintainability. As Jony Ive put it, "Simplicity isn't simply the absence of clutter."

TRUE SIMPLICITY REQUIRES:

- Careful regard placed on nuanced creative thinking around the proximity and grouping of objects.

- The creative use of progressive disclosure to guide the user through complex workflows.

- A deep understanding of what features are key to helping users achieve their goals, resulting in the elimination of nonessential features.

Remember, when someone buys your product, they are paying you to think for them.

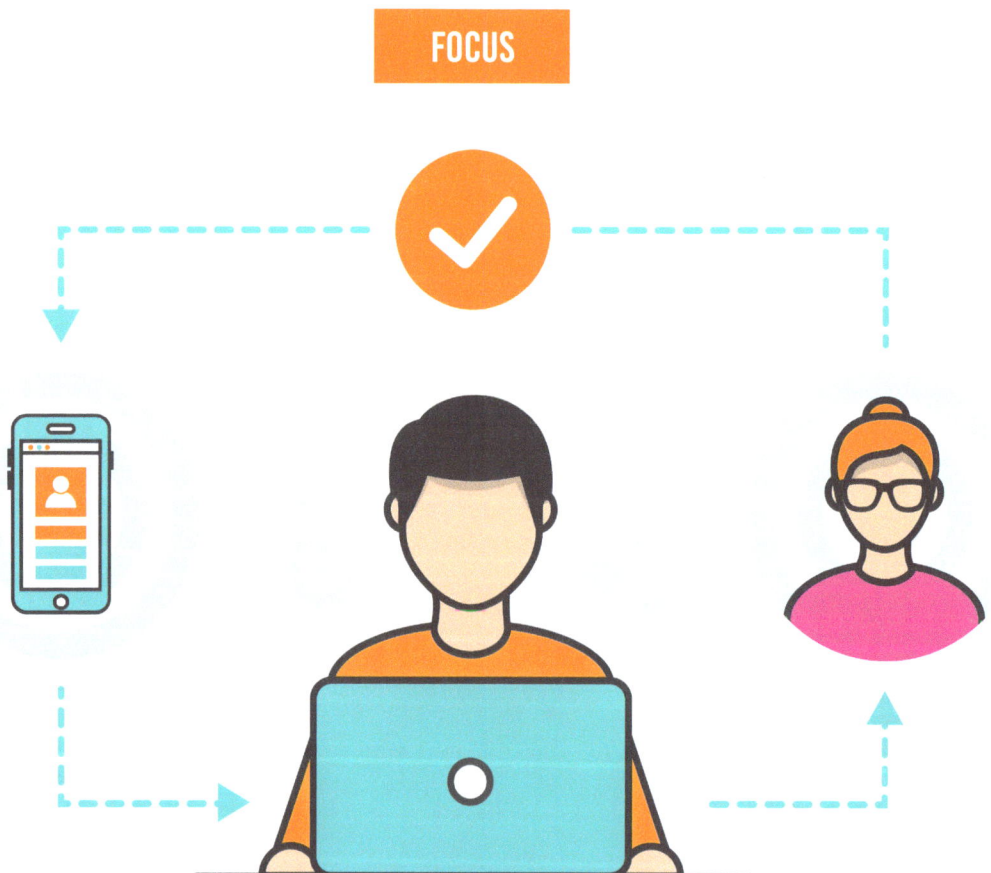

When you apply intense focus to a problem and align your product as a solution to the goals of your user, you create a stable, predictable, and enjoyable environment. When users intuitively understand the conceptual model of how your product benefits them and what it enables them to accomplish, an emotional connection is created. When you focus on the user, you provide the feedback needed to successfully use your product. Users should receive feedback throughout the experience so they know what their current status is, what's happening now, what has happened already, and what they can do in the future.

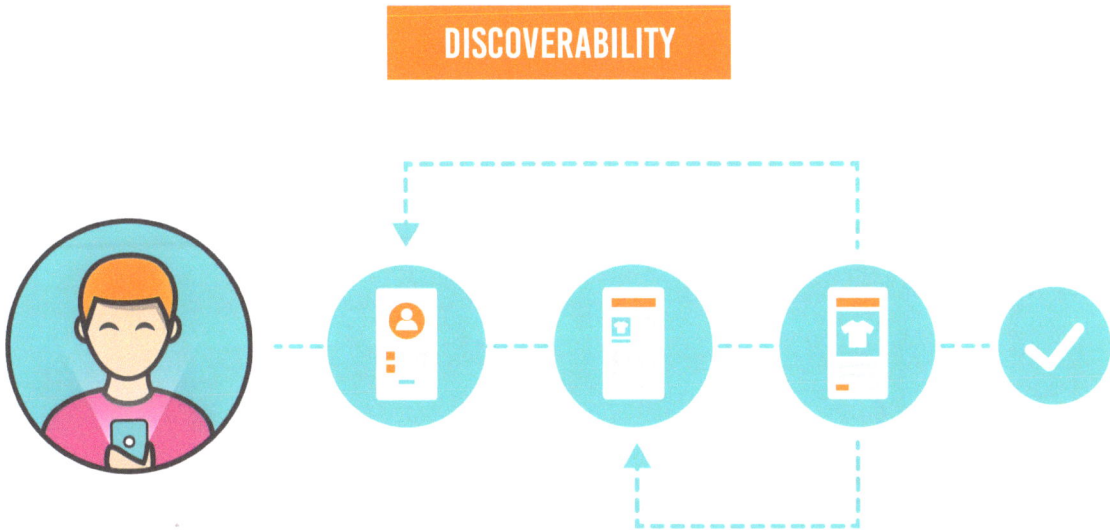

DISCOVERABILITY

Digital products should be easy to navigate, with streamlined and simple workflows. If using your product requires an operation manual the likes of which would impress IKEA, you need to rethink the experience. Users should automatically be able to navigate your product without having to be taught.

THIS EXTENDS TO THE STATE OF THE APP:

- What are the entry and exit points?
- Can the user come back and pick up where she left off?
- Does the state sync across devices or platforms?

On the other hand, a well-thought-out onboarding process can be a great add-on benefit to any app. When an app opens for the first time with a default, albeit optional, tutorial, it gives the user the opportunity to get more value out of their experience. Even though your product should be intuitive to use, a streamlined tutorial is a nice-to-have touch.

"DESIGN IS REALLY AN ACT OF COMMUNICATION, WHICH MEANS HAVING A DEEP UNDERSTANDING OF THE PERSON WITH WHOM THE DESIGNER IS COMMUNICATING." — DONALD A. NORMAN[15]

Donald Norman, author of *The Design of Everyday Things*, outlined several fundamental principles of design, which are worth mentioning here:

1. Use both knowledge in the world and knowledge in the head.

2. Simplify the structure of tasks.

3. Make things visible: bridge gulfs between Execution and Evaluation.

4. Get the mappings right.

5. Exploit the power of constraints.

6. Design for error.

7. When all else fails, standardize.

CONSISTENCY

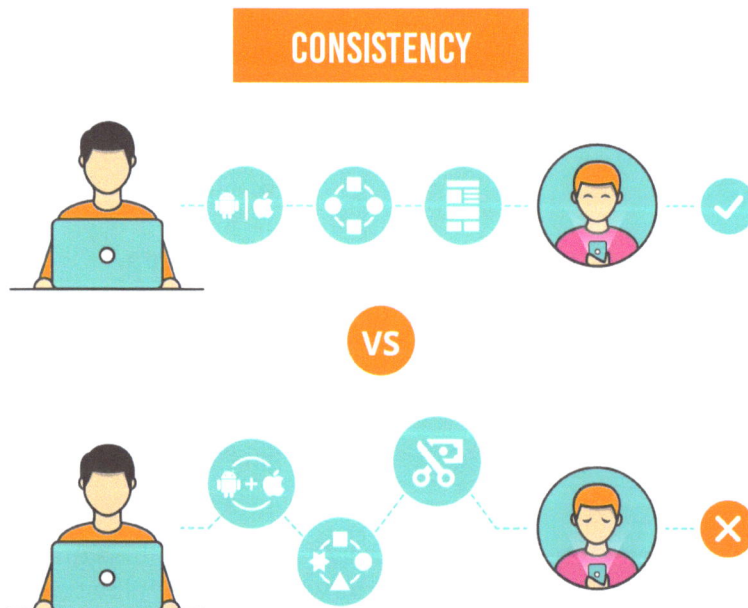

---

15 Donald A. Norman, *The Design of Everyday Things* [MIT, 1998].

Designs should conform to the platform standards users are used to — blended with your unique brand elements. This consistency improves usability. It's bad form, and confusing, to use Android design patterns in iOS and vice versa. One of the worst crimes you can commit is to use web standards when designing apps.

IT'S CRITICAL TO DESIGN TO PLATFORM.

You must also be familiar with the current design patterns of the most widely adopted apps and products. Don't let your designers fall into the trap of trying to be clever and sophisticated with revolutionary patterns. If your users have to learn a new design paradigm, it complicates the experience.

AGAIN, THE LITMUS TEST IS IF IT REQUIRES INSTRUCTIONS TO USE YOUR PRODUCT.

These principles of design are often overlooked by experienced and inexperienced product owners alike. The current business mindset is obsessed with cost savings, efficiency, and mergers, so what we see all too often is a mental model of creating Swiss-army-knife products. These products may have a gazillion features, but they don't do anything extremely well. In the product space, we've found this is a recipe for cost overruns, inflexibility, poor user experience, and ultimately a terrible product. Good design becomes nearly impossible to achieve.

Even the most talented designers need to be led by a capable product owner. A lot of designers are absorbed with the craft of their work and lose site of the purpose. They are perpetually making decisions on how a project impacts their portfolio instead of taking a step back and focusing on the product's users. It's hard for designers to connect the success of a user back to their own success. This is something to be aware of as you hone in on the experience of your product.

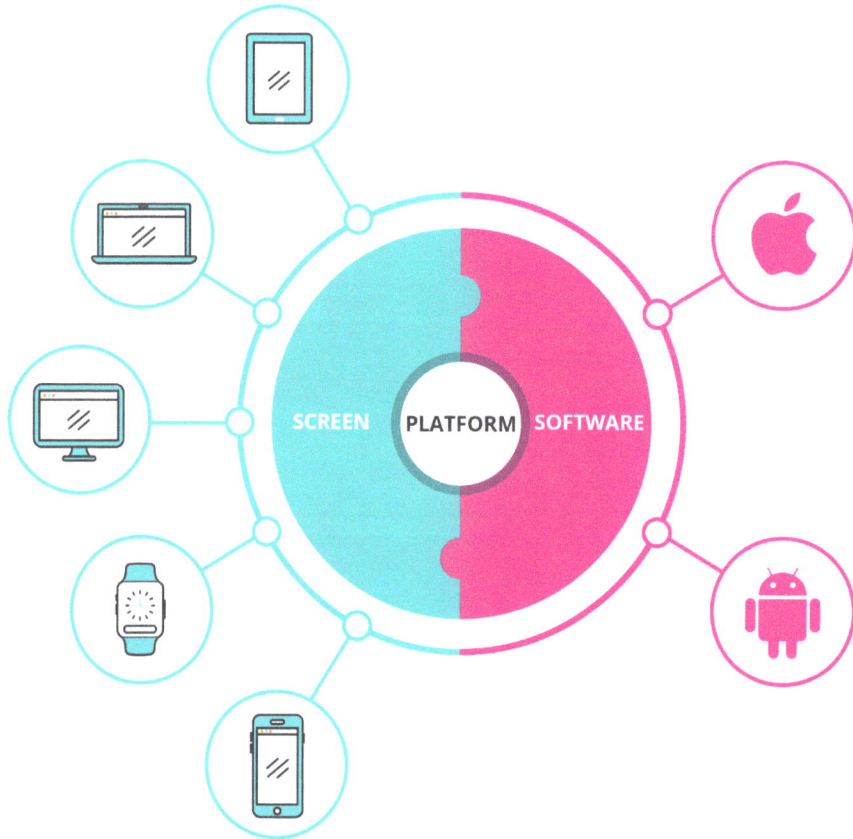

"CHANGE BEFORE YOU HAVE TO."
JACK WELCH

At this point in your development, it's time to make a rational, informed, and practical platform decision. Platform refers to both the device and the operating system your product will be built for, and might include any of the following: smartphones, tablets, the web, smart home devices, wearables like watches.

Most successful products these days are extended, connected, and synchronized across multiple device platforms, and the experience on each is customized. Gone are the days when you could just have a website. Now, you also need an app, and that app needs to work on iPhones, Androids, and all kinds of tablets. There is an ever-increasing ecosystem of smart digital platforms, and to create a successful product, it's essential to have a deep understanding of the capabilities of all of them.

STILL, YOU NEED TO START SOMEWHERE.

When you're envisioning a new product, you should carefully consider an initial platform and focus on what will work best for the user to successfully accomplish his or her goals. You might need to focus on more than one platform, but you might not need to cover every platform right away. It's so common to hear product makers say, "We need an app, too!" before they've truly solidified their product vision. This makes us cringe, because most of the time, they don't have a clue what they need, and are just trying to play catch-up or make their boss happy.

When choosing a platform or two to focus on, make purposeful decisions. Don't just pick the hot new property; build the best user experience. We recommend starting with a base platform and adapting your product to the other platforms that will add value to your users. For most people, that base platform is either going to be an app or a website. If a website, you'll need to make sure it works on every major browser. If an app, decide upfront whether to design for iOS or Android — or both.

You don't have to commit 100%. It sometimes happens that user testing results in a platform change. Although you can usually guess up front which platform is going to be best for your product, there are always edge cases where the user's environment determines which platform they tend to use. It's not always the one you expect.

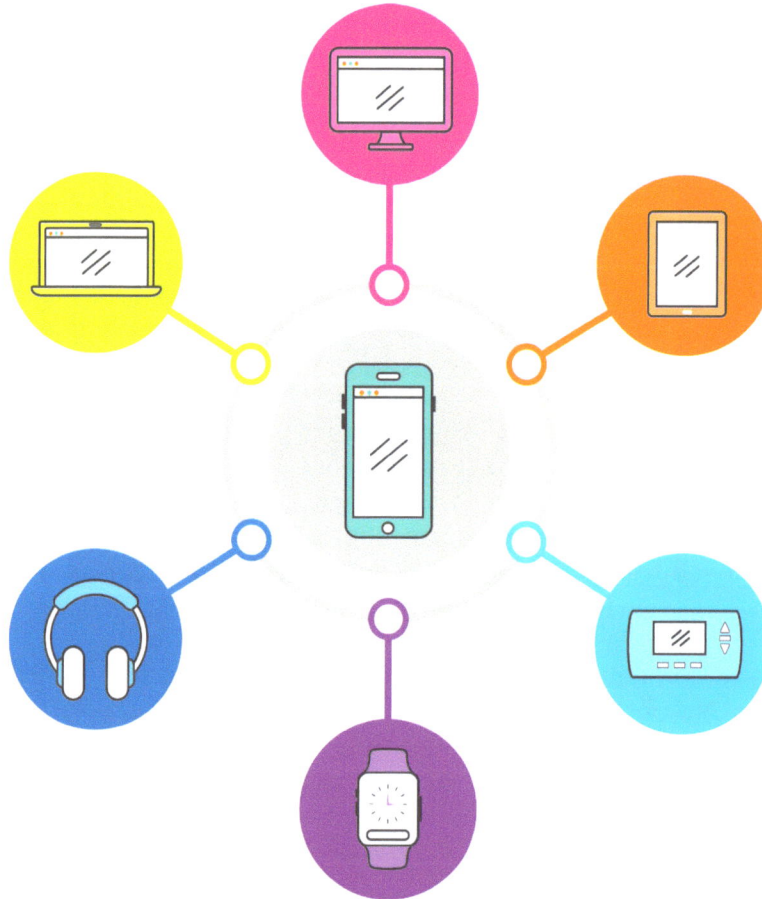

From the human perspective, mobile is the hub of it all. Most people carry their phones everywhere, and those phones are always on. They have sensors to detect just about everything, and they play a key role in most digital experiences.

If you're asking yourself, "What's the time, place, and type of environment my users will be in?" it's typical you'll come back to the phone. Regardless of what type of product you are making, there will probably be some type of mobile experience in the mix. Determining whether this experience is a mobile-responsive website or an app is largely circumstantial and capability specific, although we do highly recommend that you never try to jam your website or bloated web app onto a phone.

If you are building an app and aren't a seasoned veteran who understands when it's appropriate to deviate, stick to the native software development kits (SDKs) for each type of device. The one exception to this is with game development. There are some use cases where using a platform like Unity might be the best option for your use case.

While this book is written for anyone creating a digital product where user/customer experience is pivotal, we would be remiss here if we did not caveat this mission. There is one area of digital product invention that this book does not thoroughly cover, and that's gaming. Creating games requires virtually every skill under the sun:

MOTION GRAPHICS · ANIMATION · PSYCHOLOGY · ANTHROPOLOGY · CREATIVE STORYTELLING TECHNIQUE · CINEMATOGRAPHY · VIDEOGRAPHY · TEXTURING 2D AND 3D MODELING · APPLIED MATH · SOUND EFFECTS · COMMUNITY DEVELOPMENT · AND SO MUCH MORE

We highly recommend starting with the book The Art of Game Design: A Book of Lenses by Jesse Schell if you're interested in launching a gaming product or company.

Assuming mobile is the core of your product, we recommend creating a platform map where you chart all key product categories in the digital space from desktop computers to wearables (like watches) to smart home devices (like Amazon Echo). The key here is to consider the user goals and map those to platforms.

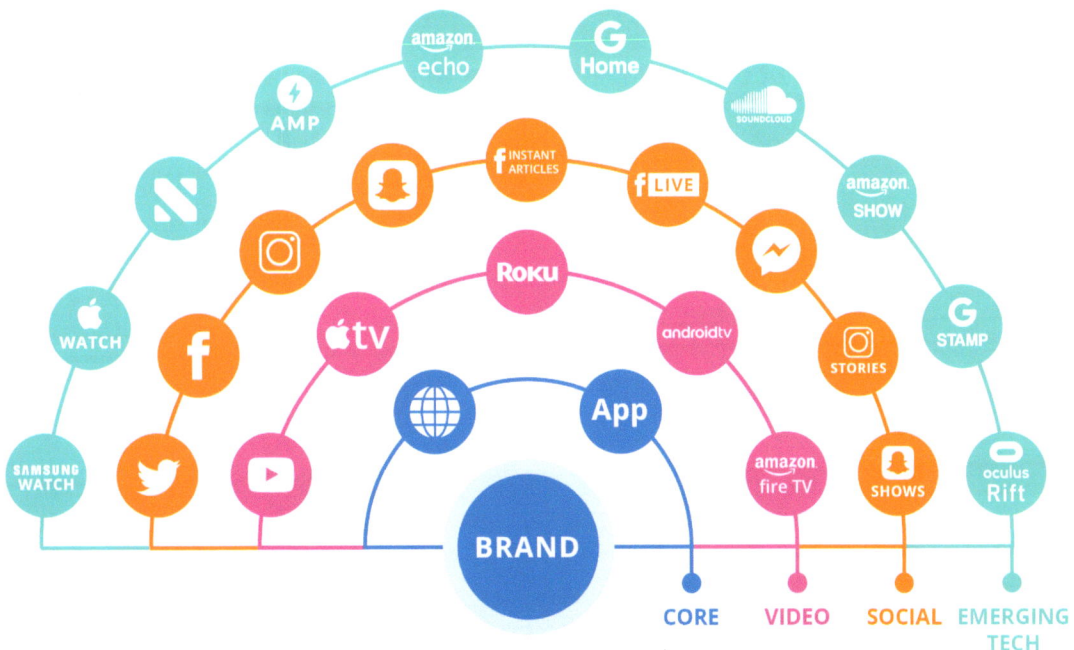

CORE   VIDEO   SOCIAL   EMERGING TECH

For example, if you are developing a business application, you can assume a few things about its use:

1. It's likely to be used in a home or office setting.

2. It's likely to involve working for extended periods of time.

3. It's likely to involve multitasking and require lots of screen real estate.

Knowing all this, it's obvious that a desktop computer with a large screen and plenty of CPU horsepower would be the tool of choice. But what's the next closest platform? Well since it's a business application, and since business is so mobile these days, you'll need a mobile version too — probably a laptop or tablet, both of which have bigger screens than a smartphone. It's highly unlikely that a wearable device like a watch will be useful, unless there is some direct tie-in to the desktop experience, like a secured environment where the watch serves as part of a two-factor identification process.

A caveat to this general business application advice, though: increasingly, more and more apps are being designed for mobile workers — people who do things like appraise property, conduct door-to-door sales, or fix equipment in the field. For these types of applications, a mobile experience is typically more important than desktop.

HERE'S ANOTHER EXAMPLE:

You're developing a fitness app with the phone as the hub. However, a smartwatch is in the mix as well, because it creates a more efficient, personalized experience for the user while working out. Connecting the phone and watch experience is key.

You might be reading this all and thinking, who would ever screw this decision up? It's so obvious! But not every product has such an obvious platform of choice. And we've found that a lot of product owners who don't understand their users do, in fact, screw this up.

We recently worked with a client convinced their app must work on both iOS and Android. This was a reasonable idea, except that after extensive research, we knew that more than 80% of their audience was on iOS. Their insistence on developing for Android anyway was costing them time and money, and they were budget pinchers, so it didn't quite jibe.

With their limited resources, our development was handicapped by being fragmented between the development of two versions of the app under a tight deadline. Instead of making an incredible app for iOS, we made two "good enough" versions.

There are some key questions you can ask yourself about your users to get to the heart of this decision.

WHEN AND WHERE WILL THEY BE WHEN THEY USE YOUR PRODUCT?

WHAT DEVICES DO THEY TYPICALLY HAVE WITH THEM?

WHAT INPUT FUNCTIONALITY WILL BE REQUIRED TO COMPLETE A TASK? IS IT BETTER ACCOMPLISHED WITH A TOUCHSCREEN, A PEN TABLET, OR A KEYBOARD AND MOUSE?

HOW COMPLEX IS THE TASK? DOES IT REQUIRE LOTS OF SCREEN REAL ESTATE OR SPECIALIZED TOOLS?

WILL THE USER NEED TO BE CONNECTED TO THE INTERNET WHILE USING YOUR PRODUCT?

ARE THERE ANY DEVICE-SPECIFIC CAPABILITIES THAT THE PRODUCT IS CONTINGENT UPON? (THINK THINGS LIKE BIOMETRIC SENSORS OR CAMERAS.)

WILL THE SCREEN BE VIEWED IN A SHARED ENVIRONMENT? DOES IT REQUIRE A/V PRESENTATION CAPABILITIES?

IF YOU TALK TO USERS AND UNDERSTAND HOW THEY CAN BEST ACHIEVE THEIR GOALS, THE PLATFORMS WILL NATURALLY ALIGN.

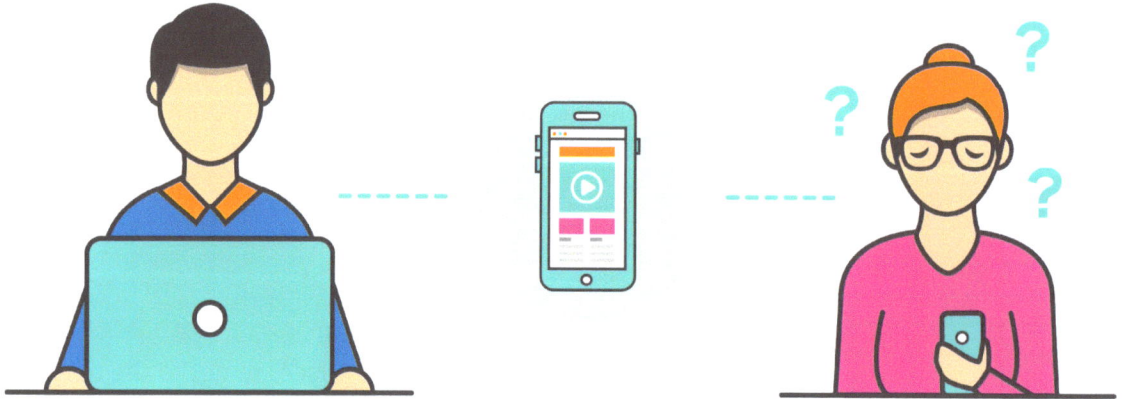

Even as communication and customer insight are skyrocketing thanks to the rise of digital technology and the shift to mobile, there is still a major gap between your perception of your customers' challenges and what they're actually concerned with. Unless you are designing a product for which you are the target market — like a product to help product-makers make products, for instance — **YOU SHOULD PROBABLY TAKE YOURSELF OUT OF THE EQUATION WHEN IT COMES TO TRYING TO UNDERSTAND USERS.**

Research shows (and common sense dictates) that product designers have more technological literacy and often even a higher IQ than the users they're designing for. Your biases and your intimate knowledge of the product and its features are probably going to adversely affect your ability to role-play as one of your users. As a product designer, what you like is not relevant. Your technology skill level is equally irrelevant. Even those on your team who are "less technical" might be more technology-savvy than the average user (see box).

THE 4 LEVELS OF TECHNOLOGY PROFICIENCY

An ambitious study on technology literacy was undertaken by the Organisation for Economic Co-operation and Development, which tested 215,942 adults across a range of ages and countries. The results partition adults into four basic levels of technology proficiency.

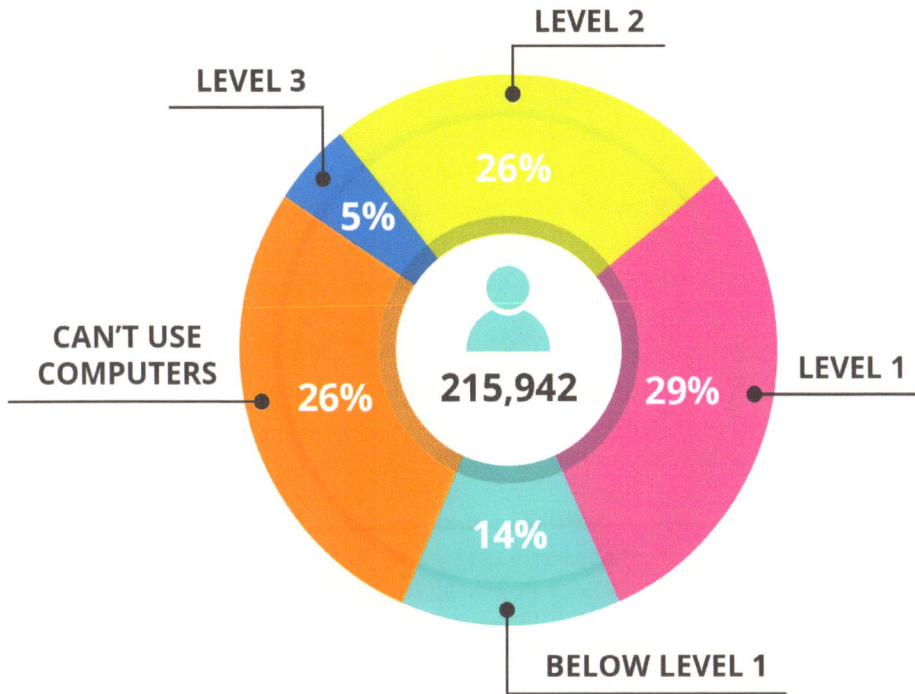

LEVEL 3

LEVEL 2

CAN'T USE COMPUTERS

215,942

LEVEL 1

5%

26%

26%

29%

14%

BELOW LEVEL 1

1. **Below Level 1** (14% of adults tested) Can conduct only the most basic tasks, like pressing a button when explicitly told to do so.

2. **Level 1** (29% of adults) Can complete tasks with familiar applications like web browsers and email software, as long as there are few steps and variables — for instance, "replying all" to an email.

3. **Level 2** (26% of adults) Can improvise with interfaces they have never seen (like a unique online form) and navigate across pages and applications to complete a task.

4. **Level 3** (5% of adults) Can complete tasks that involve navigation across pages and applications, use in-application tools (like a "sort" function), and handle multiple steps and operators; they are adept at applying inferential reasoning to solve problems.

You might notice that the percentages above do not add up to 100. That's because a fifth group of adults exists: the 26% who can't use computers at all.[16]

16 "This Chart Shows How Computer Literate Most People Are," Lifehacker, December 7, 2016, https://lifehacker.com/this-chart-shows-how-computer-literate-most-people-are-1789761598.

"ONE OF USABILITY'S MOST HARD-EARNED LESSONS IS THAT YOU ARE NOT THE USER. THIS IS WHY IT'S A DISASTER TO GUESS AT USERS' NEEDS." — JAKOB NIELSEN[17]

You are probably in the top 5 percent of adults when it comes to technology literacy. You might be designing for an elite audience, but even among that audience chances are you're toward the very top of the technology elite. Even a target demographic of highly intelligent engineers is coming to your product cold, which gives them a handicap. So if you think something is easy, self-explanatory, or a piece of cake, the point is, you can never be sure. You must always test out these assumptions.

User-centric design philosophies can be honed and refined through extensive research, and we'll talk about that in detail in the next chapter. But there are some ground-level principles that you and your UX team should know about users and the internet. You can eliminate a lot of failed experiments early on by ensuring that you as the visionary keep in mind the overall technology literacy of nearly all internet users. This is why we start with a research-first approach, so you can avoid the pitfall of guessing what your audience needs.

If your product is geared toward a broad consumer audience, you'll need to simplify your user interface (UI) as much as possible. Assume that many of your users will be in the Level 1 and 2 categories mentioned in the box.

IN OTHER WORDS, THEY NEED INTERFACES WITH:

- Few steps and little to no navigation required
- A minimal number of operators
- Any criteria the user needs supplied in an explicit way
- Few monitoring demands and no need to integrate information

If your design is more complicated than that, you're limiting your user base to that top 5 percent of adults who are the most computer literate. Start with a good hypothesis and a deep understanding around where your users fit in the technology landscape, and never overestimate their skill level. Always hire, test, and aim toward building products that are intuitive for users.

This entire section can be summed up by the old adage K.I.S.S.: Keep It Simple, Stupid.

---

17 Jakob Nielsen, "The Distribution of Users' Computer Skills: Worse Than You Think," Nielsen Norman Group, November 13, 2016, https://www.nngroup.com/articles/computer-skill-levels/

2.0 LEAN, MEAN RESEARCH

You know you can solve a user problem. You're passionate about the project and willing to spend a lot of time working on it. You've gotten some initial feedback on your idea, and you've laid the foundation for your design. To connect these things together, it's time to do some research.

Research is where you gather data to prove that your idea has legs. You learn more about your potential market, run a feasibility audit, enlist data to build your capability, and, at the end of the day, make sure that every decision you are making aligns back to your original vision.

Research will give you a solid business case for your product. But it's easy to get carried away. Some entrepreneurs go down the rabbit hole of conducting product research and never come out. Others resurface, but with an altered mindset. They take the little "research pill" and lose sight of their original vision and commitment to solving a problem for their potential users.

---

18 Will Caldwell, "The Golden Rule of Starting Up: Product-Market Fit," Entrepreneur, December 10, 2013, https://www.entrepreneur.com/article/230287

For this reason, we always advocate doing lean, mean research. In other words, don't spend a lot of money. Don't hire an expert. Don't lock yourself up in a room with a focus group for days on end. Instead, conduct quick and dirty research using the tactics we'll describe in this chapter.

IT SHOULDN'T TAKE MORE THAN A FEW DAYS, AND AT THE OTHER END, YOU SHOULD BE READY TO DIVE RIGHT INTO CHOOSING A DEVELOPMENT MODEL.

THE 5 BENEFITS OF LEAN, MEAN RESEARCH

Taking a lean, mean approach to research has five valuable benefits:

1. IT'S EFFICIENT

Traditional research just doesn't work within the time and budget constraints of developing new applications and websites. Interviewing your own contacts and strangers on the street yourself can easily provide adequate sample sizes and insights without the formal overhead.

2. IT'S SPECIFIC

The best person to show your product to is the person you're designing it for. Find someone who fits that description. This protects the customer-centric focus of your product. You might have pet features that you think are game-changers, but quickly find out that your audience doesn't really care about them — or, worse, things they expect that you didn't think of.

3. IT'S ITERATIVE

Customer feedback should be employed early and often. The goal is to put as many prototypes between the initial idea and the product launch as possible. This allows your product to become the best version of itself through multiple quick iterations of incremental improvement.

4. IT'S ACTIONABLE

Actionable insights come from knowing not only what is important but why, and how to specifically modify the product. The goal is to understand and solicit customer feedback that leads to improvements to the product.

5. IT ENDS ARGUMENTS

Research can end arguments. Perhaps there's a feature of your application, or the design, or the UX, that you and your partner don't agree on. You both feel very strongly about your opinion, and no one wants to budge. With research comes a winner. Rather than arguing about it forever, put it to the absolute truth of user testing.

MARKET RESEARCH IS A STEP OF THE PRODUCT-DESIGN PROCESS YOU SIMPLY CANNOT SKIP.

It exposes holes in your assumptions. It tells you where you're missing information. It helps predict how your future users will think, feel, and act when using your product. It will help you decide whether this product is worth building before you spend thousands (or hundreds of thousands) of dollars working on a bad idea. There is no silver bullet to building successful products, but a little research comes close.

"THE MOST IMPORTANT WORD IN THE VOCABULARY OF ADVERTISING IS TEST. IF YOU PRETEST YOUR PRODUCT WITH CONSUMERS, AND PRETEST YOUR ADVERTISING, YOU WILL DO WELL IN THE MARKETPLACE."

DAVID OGILVY

If you're like most entrepreneurs, you probably feel pressure to create your product as quickly, efficiently, and cheaply as possible. Excessive research might feel laborious and unrealistic, given your time and financial constraints.

But market research is important. It will determine whether there's an actual audience out there for your product. It will help you explore various UX directions, identify the most promising ideas early on, and refine and optimize these ideas through prescriptive, actionable insight. It's where your personal passion meets the willingness of people to buy.

THE GOOD NEWS IS THAT MARKET RESEARCH DOESN'T NEED TO BE EXHAUSTIVE OR EXPENSIVE.

You don't need to hire a fancy research expert or firm to get the valuable answers you need. You don't have to log long hours at a library or in a usability lab.

INSTEAD, USE THE TOOLS ALREADY AT YOUR DISPOSAL:

- Your own network of contacts
- Man-on-the-street interviews
- Industry trade shows, where you can learn from thought leaders on the subject
- A mentor
- Google (of course)
- Fake-out ads (see box)
- A video or post on a site where your audience frequents

Of the above, the most important is to talk to actual potential users face to face. Hunt down actual people who you think would use your product, and show it to them (as we mentioned in the beginning of Chapter 1). This can be done in as little as a few hours, with a handful of participants.

THE FAKE-OUT AD

The fake-out ad is one of our favorite methods for conducting lean research early on. Thanks to the easy, cheap, instant nature of Google and Facebook advertising, you can set up online ads quickly without even having a product to send people to. By creating several versions of your ad, you can do some quick-and-dirty A/B testing on things like your product name, your product description, your basic marketing verbiage, specific target markets, and calls to action. It doesn't matter that these ads don't go anywhere. What matters is that you can see who clicked on what, with the platform's robust reporting.

Running an imaginary ad for your future, mythical product on Facebook or Google might seem a bit weird at first, but when you realize how much value you can get from this type of research, at a pretty inexpensive price point, you'll be hooked. It's one of the best ways to figure out if the product you are building is something people actually want.

TECHNICAL INSIGHT OVER MARKET RESEARCH

"TECHNICAL INSIGHTS CAN GIVE COMPANIES AN UNFAIR ADVANTAGE, WHICH IS WHAT MAKES THEM SO IMPORTANT." — ASHMEET SIDANA[19]

Google is in a great position to rely on data for insight. Google Search conquered all prior search engines because its designers figured out a way to rank webpages in a ways that presents optimal search results. This Google tool, PageRank, works by counting the number and quality of backlinks to a webpage in order to deduce its importance.

Technical insights are at the heart of many of Google's most successful products. And for this reason, Google's core strategy is to prioritize technical insight over market research. Market research has an air of theory to it. Technical insight is tangible.

---

19 Ashmeet Sidana, "Technical Insights in Startups," *Engineering Capital*, accessed July 7, 2017, http://www.engineeringcapital.com/essays/technical-insights-in-startups/

2.2 ENTER THE ALMIGHTY USABILITY STUDY

"BUILD SOMETHING 100 PEOPLE LOVE,
NOT SOMETHING 1,000,000 PEOPLE KIND OF LIKE."
BRIAN CHESKY[20]

Lean market research will carry you far. But at some point in your product-making journey, you should invest a little bit of time into a more formal usability study. Here, in a one-on-one setting, guided by a professional researcher (that might be you or someone on your team), you can observe users as they engage with your product.

After you witness the first two or three users engaging with your product, you'll learn roughly 50 percent of what you need to know. With five users, you'll be 80 percent of the way there. And with each additional participant, you'll start to hear similar comments and themes, with a smaller number of new unique findings. By testing fifteen users, you'll uncover 99 percent of the usability problems that lie within your product.

ARE YOU REALLY MOBILE?

These days, there's no such thing as a product that doesn't need to be mobile. Your product is either designed for mobile, or it has a mobile component.

So why is mobile-experience design not a given?

One of the biggest things we see come out of usability testing is the fact that products don't display or work well across mobile devices. And usually this is because they aren't being thoroughly tested across mobile devices.

You can't just shrink down the website version and hope people have their glasses on. A squished desktop experience is virtually unreadable on mobile and often responsible for rampant bouncing off apps and sites.

Mobile-experience design doesn't just make sense for the sake of user's eyeballs; it's also a search engine ranking factor, by the way. If your website is not mobile-optimized, people might never even see it in the first place.

20 Inigo, "Brian Chesky: Airbnb cofounder," *Medium*, July 31, 2014,
      https://medium.com/founders-quotes/brian-chesky-b6a12620a26e

These numbers are not scientifically proven, but we've seen them play out with experience. Once, we needed a usability consultant on a project, and at the time Microsoft was known for its high caliber of user research, so we specifically scouted out a woman who had worked on the company's Xbox testing team. With advanced degrees in both psychology and applied mathematics, Alexa (not really her name) was good. She knew the traditional research methods well, but she also had an intuitive knack for less formal, more grassroots user testing.

Often, Alexa would have test subjects meet her at a coffee shop, and she'd bring a simple laptop. She was gifted at getting users to speak their thoughts while using products, without actually leading them. When interviewing test subjects, she would always bring in an outside observer, who might be disguised as a stranger sipping coffee. The observer would rarely interact with the user during the testing, but was Oscar-worthy at bumping into the test subjects while they were waiting or on their way out the door, to get more information from them. Our work and conversations with Alexa confirmed for us the truth in our philosophy. We were already familiar with the principle that it only takes a few users to identify usability issues, but working with this team really solidified the truth of this principle.

Since then, we have revalidated it time and time again and seen that nearly all the user-experience problems are uncovered with the first handful of users.

THE ONLY CAVEAT HERE IS THAT THE USERS HAVE TO BE TYPICAL INTENDED USERS OF THE PRODUCT.

If your product is meant for average consumers on the street, but you're using your grad-school engineering buddies to test it out, you're not going to get reliable results.

Usability studies allow you to create in-depth analysis based on watching multiple people perform the same tasks as they provide detailed commentary and opinions. By tallying the experiences of all the participants, you'll arrive at a holistic understanding of what's working and what needs improvement.

WE CAN'T STRESS ENOUGH HOW IMPORTANT IT IS TO TAKE AN ITERATIVE APPROACH TO RESEARCH.

As you conduct the various types and levels of research, always keep in mind that acting on your findings is the most important element of the research.

NECESSITY IS THE MOTHER...

Sometimes, testing the usability of your product requires creative solutions. Once, while we were envisioning a mobile-responsive app, we struggled to find a way to give users a visually accurate and also tactile experience of using the app on different devices. Screenshots on a computer screen didn't carry the full flavor of mobile responsiveness. Users couldn't see multiple options at once, and the display lacked hands-on appeal.

Our first solution was to use industrial-strength Velcro to attach multiple mobile phones and tablets to a whiteboard so that users could see and touch the interface across display types. (Actually, the Velcro turned out to be a bit too industrial, but that's a story for another time.) We then used browser syncing to display the design across devices.

This worked fairly well, but after watching a few people try to play around with the various devices, we quickly realized it's physically awkward to use a phone that's attached to a vertical, immobile whiteboard.

We needed something adjustable for different user heights and viewing angles. So we built another prototype: a Medusa-looking machine with mobile devices attached to the end of flexible arms. We called it Dr. Octopus. This became an excellent way for users to try the mobile design across various devices, and it was flexible in its usage (no pun intended). This device turned out to be so handy that we even sent two of these gadgets to agency teams we worked with frequently.

The old-school research model involved bringing an eclectic group of potential users into a room to unearth opinions, attitudes, and expectations about a product being presented. But focus groups are a handicapped concept. For one thing, as Steve Jobs would have attested to, people don't really know what they want until they have it in their hands.

Your users don't necessarily have the vision — that's your job. But they can definitely affirm and validate assumptions you have, as long as you give them a hands-on prototype to have an actual tangible experience with. In a focus group, you don't actually see the participants interacting with a prototype of your product, so you're not able to verify the experience they would have. The conversational nature of the group setting will not uncover each person's individual experience interacting with a product in any depth.

Focus groups can help you hone in on copy or run disaster checks for ad campaigns once you have your product up and running. But when you're still in the market research stage, they're not as helpful.

2.3 EXECUTE A FEASIBILITY AUDIT

With a little momentum behind your product vision, and some initial research, it's time to conduct a feasibility audit. Sometimes called a *feasibility study* or even a *recommendation report*, this audit is sort of like a disaster check. It ensures that your ideas are not wildly off base, and that you have a valid foundation for a product and business.

A feasibility audit will help you define what you need to do or not do to build your product technically, economically, organizationally, and legally. To be comprehensive, we've outlined all of the ways you can look at these components, whether or not you put together a physical document or simply ask yourself the right questions.

THE ORGANIZATIONAL PIECE

The organizational part of your feasibility audit encompasses all the practical questions about the product:

- Will it solve a real problem for users?

- How will it affect the lives of those building and using it?

- How long will it take to build, and will it still be needed at that point?

The organizational piece is about scoping out your idea and ensuring you have enough talent to pull it off in a reasonable time frame.

THE TECHNICAL PIECE

Your research should answer the following questions:

- Does the technology you need to build this product exist?

- Does the technology offer a good user experience?

- Do you have the right team or access to another team to implement this technology?

- How difficult will the product be to build?

- Do you have enough experience to build it?

To get to these answers, you can reference your product workflow and requirements in order to determine whether you have the technical expertise required, and what technical resources (hardware, software, and personnel) you'll need.

Our approach to a technical audit is generally a quick prototype of the key functionality. We want to ensure whatever the core thing/function/feature our product is intended to do is actually technically feasible. Sometimes it's possible to find another product with a similar feature that you can build off in terms of either enhancing or validating the technical feasibility.

THE LEGAL PIECE

Crucially, a feasibility audit will ensure that your product idea is, in fact, legal — an important question! For instance, does your product conflict with laws around personally identifiable information (PII), sensitive personal information (SPI), the Digital Millennium Copyright Act (DMCA), the Sarbanes-Oxley Act (SOX), the General Data Protection Regulation (GDPR), the Health Insurance Portability and Accountability Act (HIPAA), the Family Educational Rights and Privacy Act (FERPA), or the Children's Online Privacy Protection Act (COPPA)?

At this point, you might want to consult with an attorney to answer some of these questions. It's always good to ensure your product fits within existing regulatory requirements. If you decide to get funding you will need to be aware of local and national regulations.

There are some homework items you can do on your own without the aid of an attorney.

- Check the trademark database to ensure you aren't infringing on another organization's intellectual property.
- Look into whether your product will require or benefit from a patent. (Patents protect inventions; trademarks and copyrights protect intellectual property and original work.)
- Go to your local state agency to create an official business and register your business name.
- Secure a domain name or two through an accredited domain-name registrar.
- Check to see if you need any special licensing, registration, or permits, for instance with the Federal Communications Commission.

In the end, though, hiring an attorney is going to cost you far less than making a perilous legal mistake. We've made this mistake ourselves, and almost lost an extremely valuable product because of a World Intellectual Property Organization (WIPO) dispute.

THE ECONOMIC PIECE

You'll want to conduct an economic audit, also known as a cost/benefit analysis. This will help you determine the possible ROI of your product by predicting potential revenue and comparing it to costs. This is also the time to do a thorough competitive audit. If Google, Facebook, or Amazon have a product with similar features — or could build one quickly — you need to be aware of how easily they could crush you with their massive scale.

HOW TO SPY ON THE COMPETITION

As part of your research, it's smart to find out which up-and-coming companies in your arena have received venture capital (VC) backing. If your competitors are not publicly held companies, the best place to seek information on them is Crunchbase. Crunchbase keeps a comprehensive list of innovative digital companies with details on investors and funding. We also like Product Hunt and Y Combinator for information on startups and upcoming digital products.

If your idea is unique, it's founded on technical insight, and the benefits of creating the product outweigh the costs, you should build the product. It's as simple as that... assuming you don't have an even better idea.

ON BOLD NEW MARKETS

In January 2000, Steve Jobs was interviewed by Fortune Magazine on his recent reinstatement at Apple and how he planned to set the company and its product apart. This was seven years before the iPhone ever hit the market, and the iMac — remember that big, bold, Bondi-colored thing? — was the company's current hit product.

Of Apple's approach to designing hardware, Jobs said: "In most people's vocabularies, design means veneer. It's interior decorating. But to me, nothing could be further from the meaning of design. Design is the fundamental soul of a man-made creation that ends up expressing itself in successive outer layers of the product or service."

With this model iMac, Jobs decided to get rid of the fan — an element that had always been thought to be essential to the personal computer. He wanted to create a quiet machine, one that consumers had no idea was even possible. He didn't make this decision based on user demand; this was an idea Jobs dreamed up on his own. "This is what customers pay us for," he said, "to sweat all these details so it's easy and pleasant for them."[21]

While market analysis and feasibility audits are important tools, if your idea is brand new to the world, sometimes you have to prioritize your vision over everything else.

---

21 Steve Jobs, "Apple's One-Dollar-a-Year Man," *Fortune*, January 24, 2000, http://archive.fortune.com/magazines/fortune/fortune_archive/2000/01/24/272277/index.htm

If you're putting together an actual document, your feasibility audit should be lean — no more than a page or two. It's a guide, not a directive. In fact, you should be able to keep the entire shebang to just a few pages. This is an alternative to creating a traditional MBA-style business plan — not necessarily relevant in today's digital-product market.

In fact, this is the one and only time we will use the term *business plan* in this whole book. We don't really believe in those. Most of the time, a business plan just becomes a roadblock to agility because, when it comes time to react to changing market conditions, it's designed to be static. It's bad for your people and your product.

A feasibility audit, on the other hand, is more like a loose guide or disaster check. It's a blueprint to aid decision-making as well as a document you can use to guide the performance of your product and business as time goes on. It will also give you a rough understanding of the costs, resources, and time involved with getting your product to market, as well as the income you could potentially generate. And from there, you can estimate your ROI.

This is the stage of your project to look into all of the technical, economic, practical, and legal implications of building your product. A feasibility audit will help you create a schedule, and will also tell you what resources you'll need. It should never hem in your vision or hold your idea back.

IF ALL SYSTEMS ARE GO, IT'S TIME FOR THE NEXT STEP.

BUILD AN UNBREAKABLE ORGANIZATION

Any business modeled around a digital product today must excel in three disciplines:

» PRODUCT MANAGEMENT «
» EXPERIENCE AND DESIGN «
» TECHNOLOGY AND ENGINEERING «

Eventually, your company should have capable leaders for each of these disciplines, although right now, admittedly, you might be wearing all the proverbial hats. In any case, another way to think about a feasibility audit is from the perspective of these three disciplines.

You'll look at your ideas from a product management angle, to assess whether you have the vision, roadmap, and strategy lined up, and inventory crucial pieces — customer service needs, analytics, public relations, legality, financial planning.

You'll look at your product from the experience and design angle, making sure you're up to date on best practices in both design and UX, and how they will specifically affect your product — particularly if you're designing a product for a vertical that's heavily regulated or has a very specific user-interface style.

And you'll look at your project from a technology standpoint, making sure you're setting yourself up with the optimal tech stack and best engineers right from the start.

2.4 ALIGN EVERY DECISION BACK TO YOUR VISION

Most companies have about a one-year plan at best. Our experience has been that many companies don't actually have more than a couple-of-months plan. And it usually doesn't go beyond the basics: retaining and keeping clients happy, creating budgets, making payroll, and trying to find the right people to get the work you have in front of you done. This leaves managers trying to figure out what their bosses ultimately want, with little or no big-picture direction.

THE BIG PICTURE IS CRUCIAL TO YOUR VISION, BUT YOUR BIG-PICTURE PLAN DOESN'T NEED TO BE COMPLICATED — IN FACT, IT SHOULDN'T BE.

There are some simple questions you can ask yourself to clarify what you are doing and then easily communicate it to the rest of your team in five areas:

1 VISION

- What do you want to do?

2 VALUES

- What's important about it?
- What principles and beliefs guide your vision?

3 METHODS

- How will you get it done?
- What actions and steps are relevant to everyone in your company?

4 CHALLENGES

- What might stand in the way?
- Identify the problems and issues you need to overcome to achieve your vision.

5 MEASURES

- How will you know when you have arrived?
- What metrics can you use to measure success?

> ## "THE ONLY PURPOSE FOR ME IN BUILDING A COMPANY IS SO THAT IT CAN MAKE PRODUCTS.
>
> ## OF COURSE, BUILDING A VERY STRONG COMPANY AND A FOUNDATION OF TALENT AND CULTURE IS ESSENTIAL OVER THE LONG RUN TO KEEP MAKING GREAT PRODUCTS."
>
> ### STEVE JOBS[22]

By writing down the answers to these five sets of questions, and sharing them with your entire team, you will have created a detailed map for your organization and a compass to help you get there. Keeping everyone working toward the same goals, making decisions to get closer to those goals, and maintaining constant open communication, even while moving crazy fast, is your blueprint for success.

Morale is better when the team works as a collective unit toward a unified product vision. It eliminates ambiguity and uncertainty, which directly increases productivity. This becomes especially important for developers, who are often left out of discussions. Having a goal-based framework allows you to take your product in a cohesive direction in a fast-paced startup environment. It can be the glue that binds your team together.

Always make sure that goals are visible to everyone contributing to your product. This way, when it comes time to make decisions, large or small, any decision-maker within your company will have guidance.

WITHOUT THIS KEY STEP, IT IS MUCH MORE DIFFICULT TO CONSISTENTLY KEEP YOUR PRODUCT ALIGNED WITH YOUR GOALS.

---

22 Brent Schlender, "The Three Faces of Steve: In this exclusive, personal conversation, Apple's CEO reflects on the turnaround, and on how a wunderkind became an old pro," *Fortune*, November 9, 1998, http://archive.fortune.com/magazines/fortune/fortune_archive/1998/11/09/250880/index.htm

2.5 USE DATA TO BUILD CAPABILITY

Data will constantly inform the process of bringing your product vision to life. This book is no exception; it's a form of data you can add to the mix. As you continuously gather data, use it to build capability into your product. We call this evidence-based or data-informed design and development.

The word *capable* means to be able to do something really well. When talking about your product and its features, the way to identify a capability is to answer the question "What does this allow me to do?"

For example, you might get a suggestion from an advisor to include single sign-on capability into your product. This, in turn, would lead to new features, like being able to integrate social media. Data insights (including feedback from advisors and users) lead to capabilities; capabilities lead to new features; new features lead to benefits. These benefits will make your product better and should solve a problem for your users.

> ## "CAPABILITIES ARE DISTINCT FROM FEATURES AND BENEFITS. THEY ARE AN IN BETWEEN STATE THAT CONNECTS THE FEATURE TO THE BENEFIT."
> ### IVANA TAYLOR[23]

Capabilities provide a framework in which to conduct exploratory testing. Note that there's a difference between testing capabilities and testing features. Think of it this way: you have a light in your bedroom, and it has a lightswitch. This allows you to turn the light off. The switch is a feature, but turning the light off is a capability. And there might be a better way to turn the light off than to walk across the room and flip a switch — for instance, clap clap. This innovative new feature solves a problem for some users.

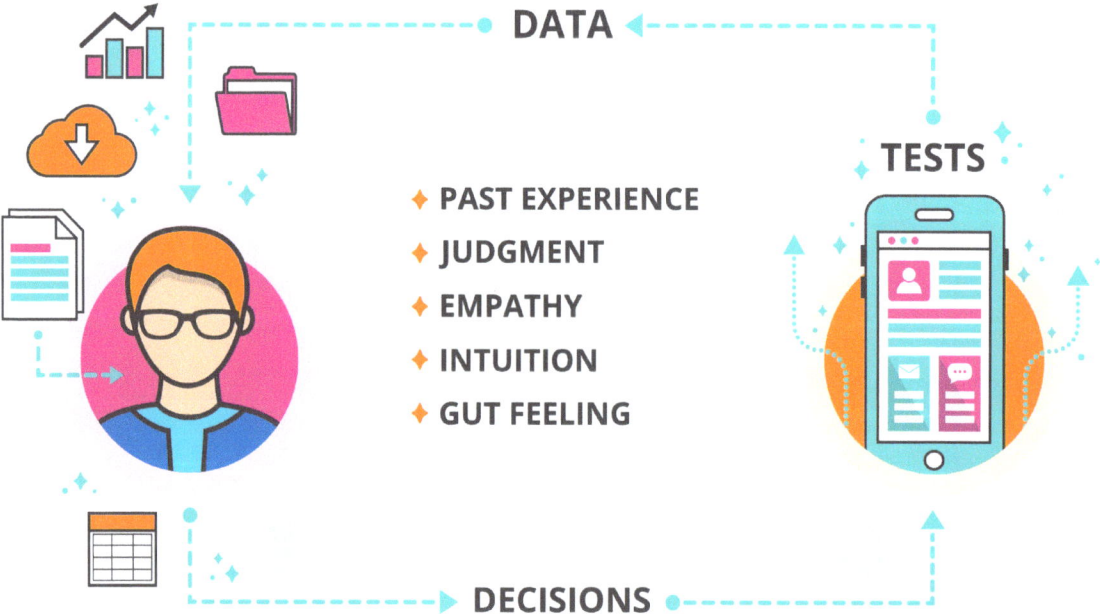

DATA

TESTS

◆ **PAST EXPERIENCE**
◆ **JUDGMENT**
◆ **EMPATHY**
◆ **INTUITION**
◆ **GUT FEELING**

DECISIONS

---

23 Ivana Taylor, "A Fresh Look at Features, Capabilities and Benefits Can Make a Sale," *American Express Open Forum*, June 29, 2011,
https://www.americanexpress.com/us/small-business/openforum/articles/a-fresh-look-at-features-capabilities-and-benefits-can-make-a-sale/

Instead of focusing on testing your features, test your capabilities. What are people trying to accomplish? Are they getting there with your product? As Gojko Adzic, strategic software delivery consultant, says, "To get to good capabilities for exploring, brainstorm what a feature allows users to do, or what it prevents them from doing."[24]

This isn't just an exercise you do once, by the way. You should always be gathering data for insights, creating capabilities, and testing them. Let data inform all of your decisions — but don't make data-driven decisions. Instead, make data-informed decisions. Data is always just a part of your decision-making process.

WITH ALL OF THIS FOUNDATIONAL WORK ACCOMPLISHED, IT'S TIME TO MAKE A VERY IMPORTANT DECISION: WHAT DEVELOPMENT MODEL YOU WILL USE TO BUILD YOUR PRODUCT.

USING DATA TO ITERATE WITH PRECISION

When it comes to digital products, data is your friend. With frequent access to quality data, you can make informed decisions rather than only relying on gut hunches.

But there is an important difference between data-informed and data-driven decision-making. Both methodologies require data, obviously. But since it is nearly impossible to acquire enough data to eliminate the human decision from the process, data-informed decision making simply makes more sense in the context of building a product.

No matter how smart an algorithm is, there is no replacement for the human mind. The human brain can make connections more quickly and astutely than computers will ever be able to — except, perhaps, when it comes to playing chess.

---

24 Gojko Adzic, "Explore Capabilities, Not Features," *Gojko.net*, March 12, 2015, https://gojko.net/2015/03/12/explore-capabilities-not-features/

3.0 TRUE NORTH: THE RIGHT DEVELOPMENT METHODOLOGY FOR YOU

Before you start to actually build your product, you'll need to settle on a product development methodology. The model you choose will inform who you hire, and vice versa. It's sort of a chicken-and-the-egg thing, so we recommend reading these next few chapters before you make any major decisions about either.

Some of the simplest-looking digital products have the most complex engineering. As you build out hundreds of thousands of lines of code, bugs and errors inevitably creep in. Catching these defects fast and early makes it less costly to repair them. When you're just starting out, you might have an elite software engineer who is able to develop code quickly with a minimal amount of bugs. This is a priceless find, by the way — but we'll get into hiring in a later chapter.

Similarly, a user-centric approach to product design means you have to be able to change features and design quickly as you get feedback. When you are small, a team of two or three people sitting next to

each other or communicating over Slack all day, it's much easier to work out the kinks on the fly. As you start to add people, teams, and communication points, being nimble becomes increasingly harder.

This is why choosing your development model early is so important. It's the foundation for how you will handle bugs and feature changes as you go, and as your team grows. Having the right model will help you catch and fix bugs and changes early, fast, and cheap. It's crucial that you choose a model that fits your team's experience, timeline, and budget.

There are more product development models to choose from than you have fingers on your hands. Each one has its strengths and weaknesses, or contingencies. It's important to match the right model to your product — and even more important to use the right model for your team. It's crucial that your team agree on a product development model, because it defines how the work will proceed and how the product will be built.

In the next few sections, we'll present the three development models that we find most common and most useful:

1. WATERFALL
2. AGILE
3. RAPID APPLICATION DEVELOPMENT (RAD)

Each of these models has its own methodology for how work should be done. Ideally, you'll choose a model up front and stick with it throughout your product's development. As you onboard employees and contractors, the product development model you're using should be a key consideration, because it can be a major source of uncertainty and stress for individuals who are used to one type of model and uncertain about trying another. It's very difficult to switch between models because it affects how team members work together.

If you find yourself in a situation where you absolutely must change development models, we suggest doing so at the end of an iteration. When a model change happens, you will need to reset and kick off the product development from scratch.

As you read the next few sections about Waterfall, Agile, and RAD, hold a solid understanding of your particular project's needs in the background of your mind, so you can investigate which model will work for you. At the end of this chapter, we'll help you make a decision.

For now, remember that the baseline complexity of your project revolves around having clear objectives and requirements laid out up front, and sticking with our constant recommendation to keep your product customer-centric.

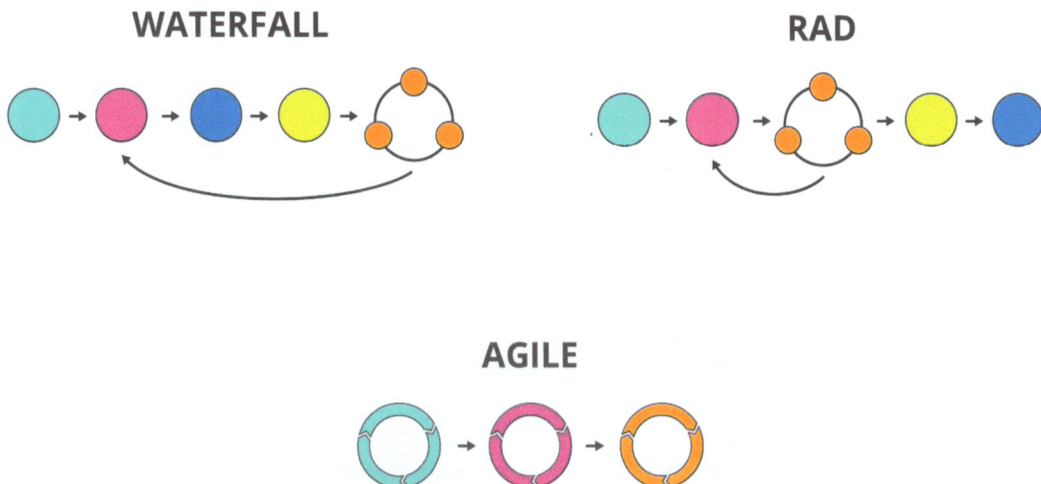

WATERFALL

RAD

AGILE

3.1 KNOW THE CASCADING COSTS OF WATERFALL

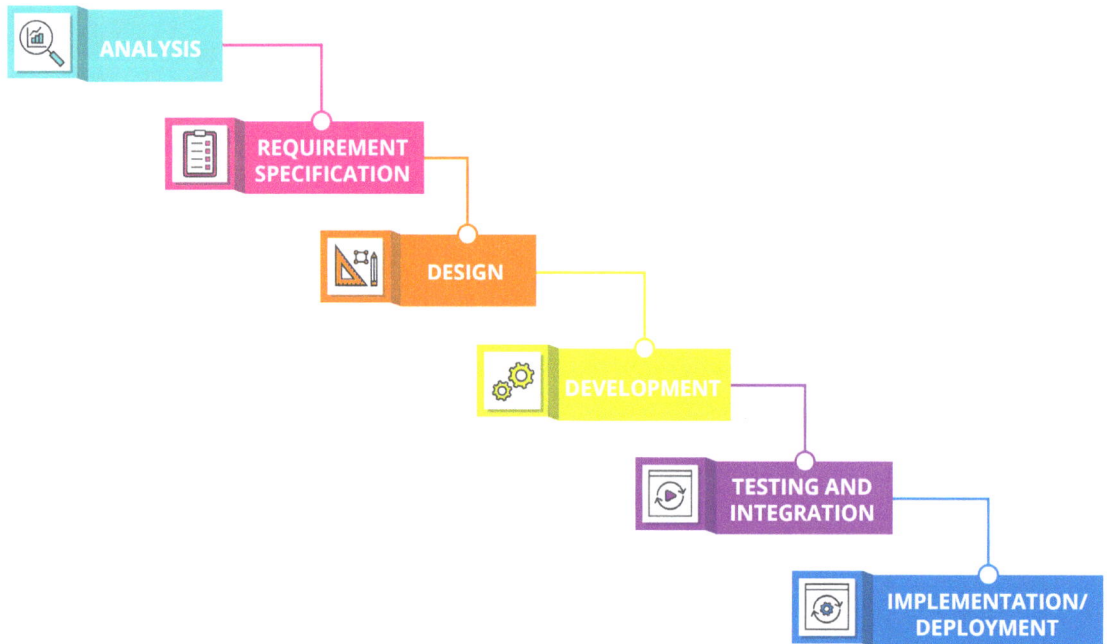

The Waterfall model has traditionally been a popular one in software development. So-called because of its sequentially flowing process — conception to initiation to analysis to design to construction to testing to implementation — it has been the most popular and prevalent product development methodology in the software product space for a while.

Waterfall is an easy model to follow. It's really not even so much a model as it is a linear system of getting things done in a way that makes logical and intuitive sense. It's easy to get teams on board even if they have diverse backgrounds, because everyone knows exactly what's expected of them far in advance. Waterfall requires little flexibility, which works well for rigid teams who don't like being flexible.

But (and there's a BIG BUT) Waterfall is losing traction and even becoming "the uncool kid on the block" as newer models like Agile and RAD are proving themselves ultimately more effective, in most cases. With Waterfall, most of the energy is put into understanding and eliminating risk through trying to anticipate the end state of the product far in advance. This model relies on extensive planning up front. If you do that planning, and follow the methodology precisely, you will create a very clear picture of the end state of your product before development even begins. In many larger organizations, in fact, the pre-planning is likely taking place years in advance, and roadmaps are developed without much understanding of how the marketing, product, or landscape might change before the product is even ready for market.

If you use the Waterfall methodology on small projects, it's like you're swimming in calm waters with gentle shifts from pool to pool. If you make a mistake it's not that hard to get back up to where you were, even with the water pushing against you. But with a bigger project, you find yourself in a much wilder and more aggressive environment. It's like a Class V rapid where any misstep might send you hurdling over a dangerous waterfall that you can't recover from. It's nearly impossible to navigate back to where you were without incurring significant effort, time, cost, and diversions. And one tiny mistake on a Class-V rapid can be fatal; the same goes for large projects relying on the Waterfall model.

ANOTHER ANALOGY WE LIKE TO USE IS THAT OF CLUSTER BOMBS.

The reason cluster bombs are frowned upon and why most countries have agreed to stop using them is that they are a threat to innocent civilians. Waterfall projects have the same effect — they can demoralize marketing, sales, and customer-service teams. And just as cluster bombs can affect a wide area and leave behind unexploded bomblets that threaten to detonate later, Waterfall projects can do essentially the same thing to your company.

WHEN WATERFALL GOES SOUTH

We once worked with a client whose web-service project was being built using the Waterfall methodology. Several of the published features had been planned out well over a year in advance, and as we became involved in the project, we quickly realized that these features were not solving problems for actual users. We tried to insert some user testing into the project and conduct a post-mortem on the current release, to no avail. The project management team was afraid of what the results would show, and wouldn't get on board. The general corporate mentality was that it's better (for both careers and departments) not to know bad news. This mindset, unfortunately, is common.

The client pushed forward with their product plan, using Waterfall, and delivering on time with all of their sprints. They were doing a perfect job of sticking to the initial plan, and constantly congratulated each other. To this day, they still see this project as a success, because they hit all their deadlines, managed the process to a T, and stayed on budget (albeit a budget of millions).

But even as deadlines were met, budgets honored, features launched, and fanfare gathered, not one single person on the 20+-person team bothered to notice whether the new features were being used by actual people. Our own analysis showed that for one main feature, which had consumed at least a quarter of the budget, the usage numbers were dismal. Less than 10 percent of users were even using it — although those numbers were being artificially driven up by members of the product team showing it off.

As user-centric product developers, this made us sick to our stomachs. We believe in creating products people want to use. This team was only interested in enhancing their visibility and clout within their organizational structure. And they've continued to "achieve" this way, moving from one roadmap to the next without missing a beat. In the time since, everyone has forgotten about that feature and moved on to the next shiny new one.

The cluster bomb this company had instilled upon themselves by using the Waterfall methodology had far-reaching effects. As they focused on the development of flashy eye-candy features, they ignored opportunities to improve upon core features that people were actually using. These oversights directly contributed to the demise of several core product features because of neglect and inflexibility.

GIVEN THE CURRENT PACE OF THE MOBILE, SOCIAL, DIGITAL SPACE, PRIORITIZING SPEED AND ADAPTING QUICKLY IN EVERYTHING YOU DO IS ESSENTIAL FOR THE SURVIVAL AND SUCCESS OF YOUR PRODUCT AND COMPANY.

Not only does your application itself need to perform quickly, but you need speed in your decision-making and your ability to change course when necessary. Waterfall's core tenets negate this approach and prioritizes a slow, careful, meticulously planned approach — which is not aligned with solving user problems. Waterfall lends itself well to fancy PowerPoint presentations in front of core stakeholders gathered in a boardroom. But it doesn't have much to do with actual users.

"'PREVENTION IS BETTER THAN CURE' APPLIES TO DEFECTS IN THE SOFTWARE DEVELOPMENT LIFE CYCLE AS WELL AS ILLNESSES IN MEDICAL SCIENCE."
MUKESH SONI[25]

If usability is the goal of iteration — which it absolutely is — then the hidden cost of Waterfall is clear. It's a fact that it's less expensive to fix a fundamental software problem early on in the development cycle.

WATERFALL

Think about a house that you are building: it's much easier to move the placement of a room or the shape of the house during the design stage. Once the foundation is poured and the structure starts to go up, this type of change gets more and more expensive.

IBM's Systems Sciences Institute reports that the cost to fix an error in the maintenance stage after product release is up to 100 times more than the cost to fix it if found in the design phase.[26]

Using a development model that allows for iteration early on, and all along, is important. Unfortunately, Waterfall is not that model for a lot of people.

---

25 Mukesh Soni, "Defect Prevention: Reducing Costs and Enhancing Quality," *Six Sigma*, Accessed June 30, 2017, https://www.isixsigma.com/industries/software-it/defect-prevention-reducing-costs-and-enhancing-quality/
26 Ibid

In a large organization, it's typical to feel like you're on the precipice of disaster, in danger of falling down a huge waterfall. So when enterprises use this method to build products, it can create a culture of fear and bluster. There's a lot of big roadmap talk, but the results are often limited. The project plans make everything sound bigger and better than what's actually happening on the ground. Sometimes, the flow of the water pushes you over the precipice — a death nail to projects, which then encounter costly overruns and scope bloat that kills the product.

Because Waterfall has started to fall out of favor with innovative developers, some companies who are afraid of straying too far from what they know still stick to some variation of this method, but pretend they're embracing an Agile methodology. This just creates more problems because the mislabeled approach has the same consequences a traditional Waterfall tactic would have.

Still, the Waterfall model has its place, especially on simple projects that can be completed in less than six weeks. In these instances, Waterfall can feel like rapid prototyping (more on that in a bit), but the key difference is the product should be shipped after the six-week interval, assuming it can stand on its own and serve a purpose for a period of time before future enhancements are required.

WATERFALL STRENGTHS

- Can work well for simple projects where there is a clear solution and requirements (that have already been validated by real users)

- Ideal for teams that have little to no experience with product development

- Gives lucid steps and objectives for each step

- Gives clear insight into risk and how to avoid it

WATERFALL CONTINGENCIES

- Simplicity comes at the cost of adaptability

- The model is beholden to upfront planning and is thus rigid

- Specific steps demand specific outcomes

- Any change in scope is a change in the product roadmap, and with Waterfall, the roadmap is not flexible

- This method fails spectacularly with massive budgets, big teams, and corporate mentalities

Some projects are successful under the Waterfall model, and some teams benefit the most from using it. This model can work well if you're developing a second-generation iteration or a refactor of a product that is already in production. It's an effective model — but in our opinion, it's not usually the most efficient or adaptive one. In fact, we've heard it said that Waterfall means "death by meeting."

NEVERTHELESS, IF YOU ARE COMMITTED TO BRINGING YOUR INITIAL PLAN TO FRUITION WITHOUT A LOT OF FEEDBACK OR ANY RISK OF A PIVOT, THIS IS POSSIBLY THE PRODUCT DEVELOPMENT MODEL FOR YOU.

3.2 IS AGILE STILL RELEVANT?

"HAVING CONFERENCES ABOUT AGILITY IS NOT TOO FAR REMOVED FROM HAVING CONFERENCES ABOUT BALLET DANCING, AND FORMING AN INDUSTRY GROUP AROUND THE FOUR VALUES ALWAYS STRUCK ME AS CREATING A TRADE UNION FOR PEOPLE WHO BREATHE."
DAVE THOMAS[27]

The next most common development model we see being used today is Agile. At its core, Agile is a methodology that prioritizes individuals and interactions over processes and tools. Instead of worshiping documentation, the emphasis is on creating working software. It's a responsive, flexible, and yes, agile approach to building digital products.

---

27 Dave Thomas, "Agile Is Dead (Long Live Agility)," *Pragdave.me*, March 4, 2014, https://pragdave.me/blog/2014/03/04/time-to-kill-agile.html

To really understand the Agile methodology, you should read The Manifesto for Agile Software Development (available at Agilemanifesto.org), which was published in 2001 by a collective of 17 software developers who gathered at Snowbird Ski Resort in Utah. Dave Thomas, one of the 17 founders, has now famously spearheaded the charge that Agile is Dead (Long Live Agility). The core issue for him is that "Agile" was originally meant to be an adjective — "The Manifesto for Agile Software Development" — but has been bastardized into a noun: "The Agile Manifesto." Now people think they can tell you "how to do Agile," which isn't possible. Agile is all about agility, and it can be summed up and measured by this one question:

HOW EASY IS THIS PRODUCT TO CHANGE?

The spirit of the Manifesto was to create a methodology that would prioritize individuals and interactions over processes — an idea that was considered very rebellious at the time. While several iterative business development models already existed, this one was more than a model. It was, effectively, a movement.

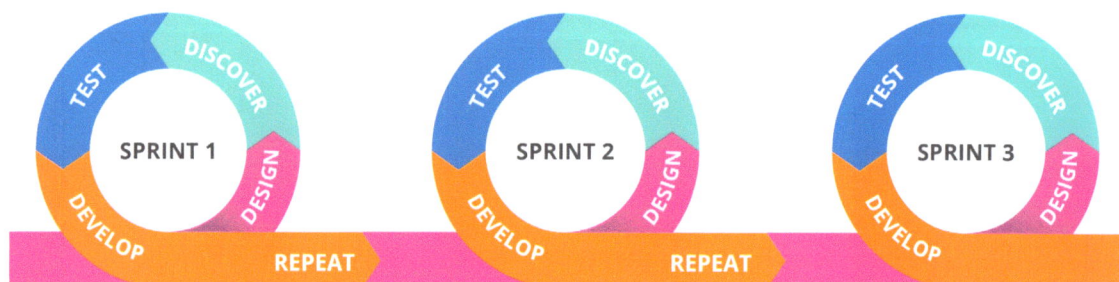

Agile is still a relevant business development model mindset, but Agile has become a software buzzword. We agree with Dave Thomas in his assertion that Agile is an adjective and should be associated with agility in the process... not a process called Agile.

We've had a lot of glimpses at a bastardized Agile process, because we work with a lot of large organizations, and they tend to have a few consistent themes:

- They're bureaucratic.
- They're obsessed with increasing productivity and efficiency.
- They fixate on the trends they read about on CMO and CTO websites.

For the last reason, bureaucrats love Agile.

In one spectacular example, a company we were working with created a massive internal marketing campaign to pitch the Agile workflow throughout the business. They created all sorts of banners and messaging about their "new, modern" development philosophy. They held all-department meetings to propagate the idea throughout the entire organization. But they didn't stop there. The leaders of this movement also swapped out the project management software and banned the use of many coding languages that didn't fit into the construct as they saw it. Consultants were brought in and employees sent out to get Agile-certified.

Ironically, productivity started to dip, so they doubled down on the Agile propaganda, trainings, and certifications. Interestingly enough, during this campaign, the actual Manifesto was never mentioned, and the principles were glossed over. The organization was obsessed with the "process of Agile," but missed the point: the mindset of agility. As things continued to spiral, the product managers became the first target. After all, the dev teams were following the Agile methodology, so it couldn't be them. They needed a scapegoat. This kicked off a period of product-management training, and the cycle continued.

The Agile Manifesto was developed by individuals who were burnt out on the traditional and cumbersome ways of developing products, and knew there was a better way to do it. But the problem wasn't the process in the first place; it was the people. Even while they moved to Agile, this organization continued to prioritize processes and tools over people and interactions. But to an outside observer, it's clear that what the company really needs is to hire the right people to better train and mentor the ones they already have, who are perfectly capable of holding the vision of agility — with the right support and leadership.

> "AT THE CORE, I BELIEVE AGILE METHODOLOGISTS ARE REALLY ABOUT 'MUSHY' STUFF — ABOUT DELIVERING GOOD PRODUCTS TO CUSTOMERS BY OPERATING IN AN ENVIRONMENT THAT DOES MORE THAN TALK ABOUT 'PEOPLE AS OUR MOST IMPORTANT ASSET' BUT ACTUALLY 'ACTS' AS IF PEOPLE WERE THE MOST IMPORTANT, AND LOSE THE WORD 'ASSET'." **JIM HIGHSMITH OF THE AGILE ALLIANCE**[28]

But changing company culture is, in many ways, much harder than changing a business development model, and this company continues to flounder to this day. When they hit roadblocks, they still seek new processes — but the problem is really the people. If they ask themselves "How easy is this product to change?" the answer is always "extremely difficult." Even though team members are following the processes and methods as described by many alleged Agile professionals, they have completely missed the overarching vision of creating agile products. They have created a complex web of interconnections that makes it nearly impossible to make changes, adjust products, and improve features — essentially killing agility.

---

28 "History: The Agile Manifesto," *Agilemanifesto.org*, accessed June 30, 2017, http://agilemanifesto.org/history.html

THE MANIFESTO FOR AGILE SOFTWARE DEVELOPMENT

1. SATISFY THE CUSTOMER THROUGH EARLY AND CONTINUOUS DELIVERY OF VALUABLE SOFTWARE.

7. WORKING SOFTWARE IS THE PRIMARY MEASURE OF PROGRESS.

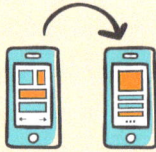

2. WELCOME CHANGING REQUIREMENTS, EVEN LATE IN DEVELOPMENT.

8. AGILE PROCESSES PROMOTE SUSTAINABLE DEVELOPMENT.

3. DELIVER WORKING SOFTWARE FREQUENTLY.

9. CONTINUOUS ATTENTION TO TECHNICAL EXCELLENCE AND GOOD DESIGN ENHANCES AGILITY.

4. BUSINESS PEOPLE AND DEVELOPERS WORK TOGETHER DAILY THROUGHOUT THE PROJECT.

10. SIMPLICITY— THE ART OF MAXIMIZING THE AMOUNT OF WORK NOT DONE— IS ESSENTIAL.

5. BUILD PROJECTS AROUND MOTIVATED INDIVIDUALS.

11. THE BEST ARCHITECTURES, REQUIREMENTS, AND DESIGNS EMERGE FROM SELF- ORGANIZING TEAMS.

6. PRIORITIZE FACE- TO- FACE CONVERSATION.

12. AT REGULAR INTERVALS, THE TEAM REFLECTS ON HOW TO BECOME MORE EFFECTIVE, THEN TUNES BEHAVIOR ACCORDINGLY.

A lot of people think they're doing Agile, but they're actually doing Waterfall. This is most often because of organizational structure. Even with the best of intentions, a larger team with a complex organizational structure can crush a true Agile workflow. As a team gets bigger, you create a "people problem": it's human nature to take the path of least resistance, and Waterfall is a safe course of action for most people. Large organizations that tend to operate from a fear-based mentality encourage employees to gravitate to what they perceive as the most secure workflow — at the expense of innovation.

One surefire way to gauge if you've got a dysfunctional Agile-Waterfall hybrid methodology is if your organization is planning product roadmaps six months or further in advance. If you are doing this, then you are doing Waterfall. Agile by its very nature should be flexible to the needs of your users. In an Agile environment, if a product manager identifies an opportunity to solve a problem, he or she should be able to prioritize and implement it sans fear of messing with the original roadmap and having to go through extensive bureaucratic approval processes.

We really like the Agile methodology, but we struggle with it. We've seen a lot of enterprising people such as software developers and business developers dilute the Agile methodology so that it's far less effective. And when put in place by large teams and massive information systems, it can be difficult to manage (as in the example we just gave).

But Agile's potential is high. It takes the better attributes of the Waterfall model, particularly the idea of starting a project with a strong plan around an end-state goal, and then adds in user feedback and the ability to change accordingly throughout the development process. Unlike the Waterfall model, it's not an automated linear process. It requires input and effort to manage.

AGILE STRENGTHS

- Starts with a vision and a plan, but is adaptable — requirements and solutions can evolve based on feedback

- Works extremely well when you know some or many of your requirements and have a list of priorities, but do not know what the exact solution will be and are willing to change those priorities frequently (think about a Kanban board, a workflow tool that can be as simple as sticky notes on a whiteboard, which can be moved around, removed, and replaced as needed)

- Rapid, flexible response to change encouraged

- Built-in mechanisms for gathering user feedback and iterating accordingly

AGILE CONTINGENCIES

- Requires a lot of communication, so if your team or your collaboration tools are not structured to support constant communication, this can be an issue

- Along the same lines, if your team is spread out over different buildings or geographic locations, Agile becomes much less agile

- It requires a good quality assurance process along the way so that iterations build on each other productively

The Agile method works best within a streamlined organization overseen by an actively involved product owner. Within a matrix organization with a "design by committee" style, the ability to make decisions within short time frames becomes handicapped and the process will likely fail.

If you're going to use the Agile methodology, we highly encourage you to use it as it was intended in the original Manifesto. If you already have a team that's comfortable using the Agile model, you're off to a solid start. It will enable you to begin product development with little planning.

"YOU HAVE TO UNDERSTAND WHAT IT IS THAT YOU ARE BETTER AT THAN ANYBODY ELSE AND MERCILESSLY FOCUS YOUR EFFORTS ON IT."
ANDY GROVE[29]

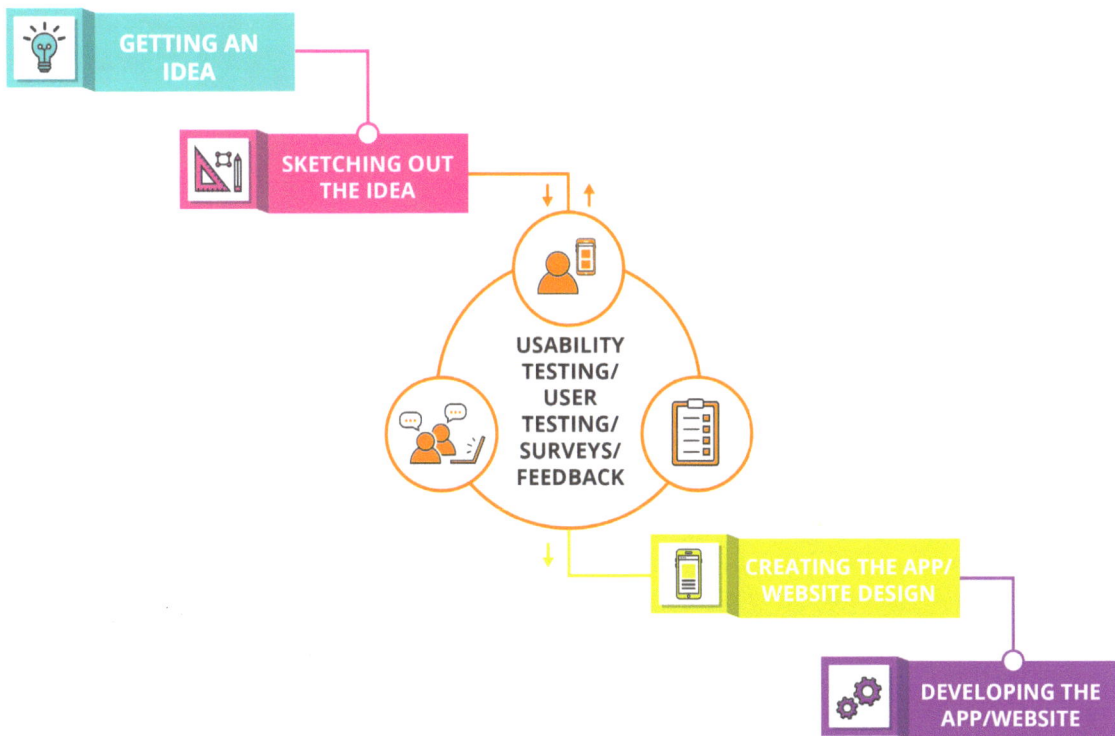

Probably our favorite development model and methodology is the aptly named Rapid Application Development (RAD), also sometimes called Rapid Prototyping. Rather than planning everything out in advance, the RAD methodology puts the focus on user experience and a streamlined process, with the assumption that a lean process will lead to solving problems quicker and, ultimately, create a better product.

---

29 George J. Church, "We're No. 1, and It Hurts," *Time*, June 24, 2001, http://content.time.com/time/magazine/article/0,9171,163135,00.html

RAD is a huge leap for many people. It turns "the old ways" of Agile and Waterfall upside down. The purpose of RAD is to maximize the rate of learning on a difficult problem or challenge by minimizing the time to try new ideas. The focus is on prototyping an experience, business model, or a product.

There are dozens of examples online about how rapid prototyping has been used to solve difficult challenges in a very short period of time.

FOR INSTANCE, GUESS HOW LONG IT TOOK TO PROTOTYPE GOOGLE GLASS? ONE DAY.[30]

LISTEN CLOSELY TO RAD

One of our favorite examples, though, is a personal one we experienced around a product we were challenged to create. Back in the days before MailChimp and HubSpot, we were enlisted to come up with a sophisticated, personalized automated email system that would leverage the client's vast amount of customer purchase data, call center interactions, and any other data we could manage to join into the system. At the time, this type of software wasn't common like it is today, and the challenge would have crushed a lot of teams. There wasn't a clear model to mimic, and even today, a project like this would typically require a big team of developers and some serious funding.

We didn't have access to those kinds of resources. We had ourselves (and this wasn't our only project, by the way). We also had one full-time developer. And we had a goal: the CEO wanted us to get the first email "blast" out in a few short weeks. Looking back, he was no doubt bluffing. He probably didn't think we could do it, and was trying to light a fire under us. But for that, we are grateful, because it forced us into a rapid prototyping mentality that actually led to success.

Given a nearly impossible task, we quickly mapped out all of the key customer data points we were aware of and then decided which would be most important based on past sales-data research. Next, we determined who we wanted to target, and spent a handful of days researching mail services, current products, demo servers, etc. Inadvertently, we were using the advice often attributed to Abraham Lincoln: "If I had five minutes to chop down a tree, I'd spend the first three sharpening my axe."

---

30 Russell Holly, "Google Glass wasn't built in a day, but the first prototype was," Geek.com, April 15, 2013, https://www.geek.com/mobile/google-glass-wasnt-built-in-a-day-but-the-first-prototype-was-1552092.

We ended up hitting our deadline, and a small email blast went out to a high-value customer segment. The response was incredible, and we immediately added a designer, a writer, and a part-time analyst to our little team. Staying with the RAD model, our new goal became to add segments and new release features to the product weekly. This was ambitious, but the impact on the product was incredible. And for us, as product makers, it was a mindset-defining experience, where we realized the power of the RAD process and the equation that looks like this:

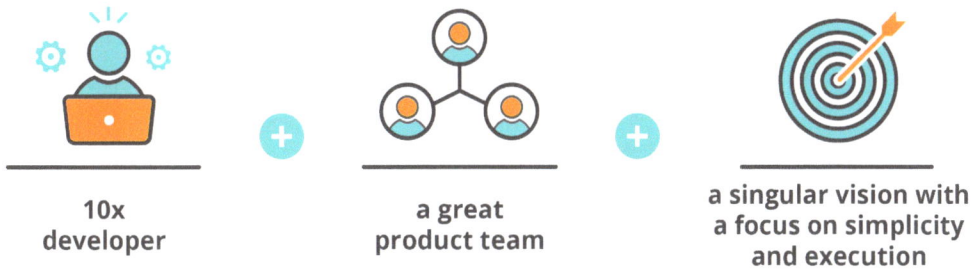

| 10x developer | + | a great product team | + | a singular vision with a focus on simplicity and execution |

Within the first six months, the email marketing product became a major driver of retention and revenue for this company. It was progressively built out to enable manual and automated campaigns, and even though it was a customized product for a specific business purpose, the functionality within that first year was superior to most of the leading out-of-the-box tools that exist now. And it became an extension of our CRM data, which was also rare and not easily available at that time. Now, tools like Salesforce make these types of integrations fairly seamless.

A nimble approach throughout the product lifecycle really works. In this example, it worked from the first release through countless iterations until we had a complex product. Another huge advantage of the RAD approach was that it forced us to modularize this product. The backend could be built in small pieces, which afforded us a huge amount of technical flexibility and the agility to evolve the product quickly as we learned what worked and what didn't. This is only one of our personal RAD use cases.

IN OUR EXPERIENCE, THIS DEVELOPMENT MODEL IS THE MOST
COST-EFFECTIVE AND TIME EFFICIENT, AND ULTIMATELY RESULTS
IN THE BEST PRODUCTS — NOT TO MENTION, THE HAPPIEST
BOSSES WITH THE MOST CHEERFUL P&L SHEETS.

Google uses the RAD methodology, and they've laid out some ground rules for rapid prototyping:

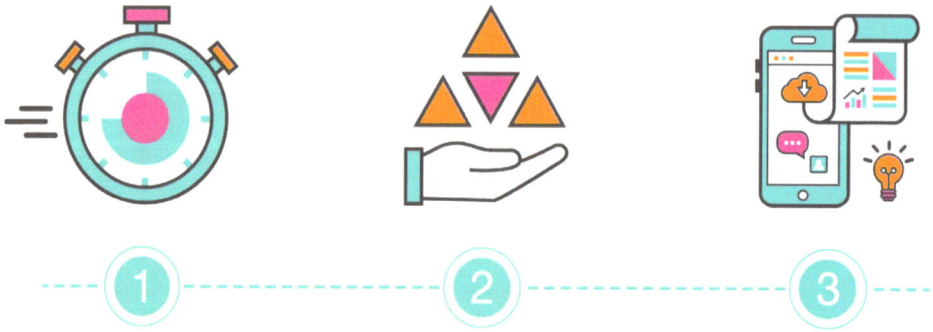

Prototyping Rule #1: Find the quickest path to experience.

Prototyping Rule #2: Doing is the best kind of thinking.

Prototyping Rule #3: Use materials that move at the speed of thought to maximize your rate of learning.

Go to YouTube and look up "Fast solutions for a brighter future - rapid prototyping entrepreneurship: Tom Chi at TEDxKyoto 2013" to watch a 14-minute video about the RAD methodology with Tom Chi, founder of Prototype Thinking and former Head of Experience at Google X. Another worthy video is Rapid Prototyping 1 of 3: Sketching & Paper Prototyping.

By the way, RAD is not specific to software development. It's a methodology you can apply to everything you do — a way of life, even. It can be used by entrepreneurs, companies, developers, students, freelancers, UX designers, project managers, and even couch potatoes. If you think about it, channel surfing is the ultimate RAD activity. We don't sit there and deliberate whether to change the channel, do we?

But channel surfing isn't necessarily a productive activity, unfortunately. The RAD methodology is meant to help you build things. Building anything new can be time consuming and cost intensive. Often, the end product is not what you originally set out to make, or it doesn't resonate with the intended audience. Following the RAD process, you focus on solving problems for end users through multiple evolving prototypes so that their experience does not get lost. Showing these prototypes to users, getting their feedback, and iterating based on that feedback leads to better products, models, and businesses quicker.

RAD, like every model, has strengths and its contingencies.

RAD STRENGTHS

- Increases rate of learning
- Reduces cost of minimal viable product (MVP)
- Reduces development time, and time is money
- Doesn't require developers (anyone can do it!)
- Makes it easy to show potential stakeholders
- Kills bad ideas (faster)
- Encourages user feedback
- Allows for quick directional changes based on that feedback
- Identifies features to build (think K.I.S.S. — Keep It Simple, Stupid)
- Works well when you don't have clear requirements and a clear solution to a customer problem

RAD CONTINGENCIES

- RAD is geared toward innovation. There are some use cases in which an established system or process already works, and therefore RAD isn't the best fit. But before you rule it out, ask yourself — does your established process really work? Or are you just attached to it?
- In a corporate environment, with a product in maintenance mode, RAD is not recommended — although, if any new features are being developed, then RAD is a good option to identify and build those new features.
- RAD is not for people hell-bent on building out a bad idea. If you are committed to your feature set, and think you know your users better than they know themselves, then stick to Waterfall!

We believe RAD is the best model for anyone who is trying to be innovative and disruptive with their ideas, regardless of whether you are an entrepreneur or an intrapreneur — someone working innovatively within a larger company. RAD enables you to start quickly proving your ideas. Rapid prototyping enables you to fail fast — and failing means learning. Learning makes your product better.

The Power of the Iterative Process

- Scott Waddell -

While working as a defense contractor, I was tasked with finding a better way to administrate an extremely complex U.S. Air Force weapons systems management interface. When I came into the project, the weapons system required administrators to know command-line Unix and repeat the same task on several different systems. It was clunky and inefficient at best.

But because the system was very expensive (like, billions of dollars expensive), not to mention mission critical (like, lives on the line critical), change was almost impossible to make. I didn't have a lot of buy-in or even the tools I needed to streamline the administration of the system. This was a classified system, so I couldn't just write code in my preferred language, Python — at least not off the bat.

At first, my goal was simply to prove to the government client that there was a problem and a possible solution. I built a shell script with some rough functionality, which proved that we could cut the administrative workload by 90%. This got me buy-in.

Next, we needed a budget. So I built a new prototype, this one with a graphical user interface (GUI), coded in a language the client approved: Perl. The government client loved this version so much that my boss got promoted because of it.

With the funding in place, my company was hired to build out the next prototype, which was eventually rolled out across the U.S. Air Force. And with this success, I was eventually given clearance to rewrite the product in Python with a Tkinter GUI. After this, we rolled it out around the world. It was a resounding success.

Without the ability to work in an iterative way thanks to RAD, this product would never have been built, never mind been successful.

3.4 CHOOSE THE METHODOLOGY FOR YOU...
AS LONG AS IT'S RAD

"GREAT EXECUTION IS AT LEAST 10 TIMES MORE IMPORTANT AND A HUNDRED TIMES HARDER THAN A GOOD IDEA." — **SAM ALTMAN**[31]

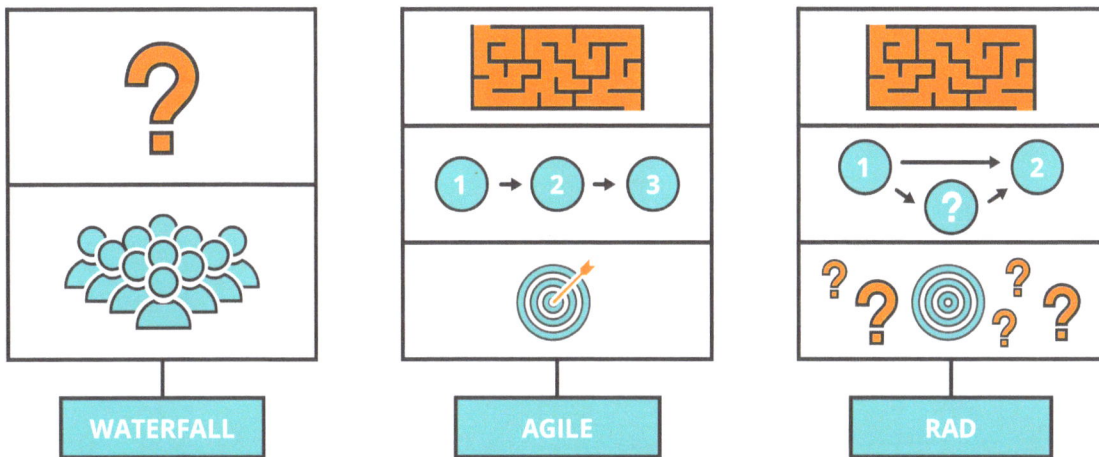

| WATERFALL | AGILE | RAD |

Now that you've read about the three main development models, do you have a better idea of which would work for your project? No? Well, we're here to help. Here are some things to think about.

If you have more than eight people on your team, more than three layers of approval, and users actively using your product, you'll probably be most successful with the Waterfall model. This model is most conducive to the high social-interaction cost associated with the size of your team, and also works best with enterprise teams that have multiple approval layers. Enterprise environments are notorious for heavy planning — for good reason; they have the resources to overcome risk.

---

31 Rajen Sanghvi, "58 Quote from Sam Altman on Startup Ideas," *Medium*, September 24, 2014, https://medium.com/how-to-start-a-startup/58-quotes-from-sam-altman-on-startup-ideas-e3582361cd4f

"I HAVE NO QUESTION THAT WHEN YOU HAVE A TEAM, THE POSSIBILITY EXISTS THAT IT WILL GENERATE MAGIC, PRODUCING SOMETHING EXTRAORDINARY, A COLLECTIVE CREATION OF PREVIOUSLY UNIMAGINED QUALITY OR BEAUTY.

BUT DON'T COUNT ON IT. RESEARCH CONSISTENTLY SHOWS THAT TEAMS UNDERPERFORM, DESPITE ALL THE EXTRA RESOURCES THEY HAVE."

J. RICHARD HACKMAN[32]

A lot of people will say that 12 is the maximum number of people on a "small" team. We like to think even smaller: 8. In our experience, anything over 8 starts to become a problem when you're in startup mode.

Harvard psychologist J. Richard Hackman spent his academic career studying the concept of teams. His conclusion? The bigger they get, the less they work. He is credited with the research that proves that as a team gets bigger, more and more links between team members must be managed, and that's where trouble brews.

- A small startup of 7 people has 21 connection points to maintain.

- A group of 9 has 36 connection points.

- A group of 12 has 66 connection points.

- A group of 60 has 1,770 connection points.[33]

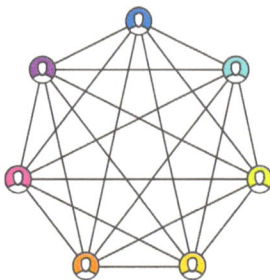

7 PEOPLE, 21 CONNECTIONS   9 PEOPLE, 36 CONNECTIONS   12 PEOPLE, 66 CONNECTIONS

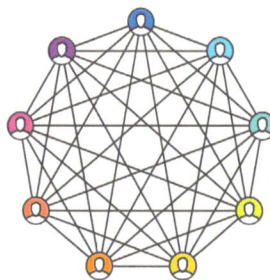

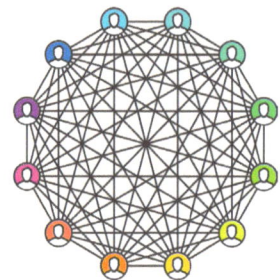

32 Diane Coutu, "Why Teams Don't Work," *Harvard Business Review*, May 2009, https://hbr.org/2009/05/why-teams-dont-work

33 Janet Choi, "The Science Behind Why Small Teams Work More Productively: Jeff Bezos' 2 Pizza Rule," *Buffer Social*, July 29, 2013, https://blog.bufferapp.com/small-teams-why-startups-often-win-against-google-and-facebook-the-science-behind-why-smaller-teams-get-more-done

For each successive member of a team you add, you increase productivity at a decreasing rate, and you increase the potential for problems — especially communication problems. We'll get into this in more detail in Chapter 4.

However, if you're used to using Waterfall, beware the trap of staying with something just because it's comfortable. Just because you've always done it, doesn't mean it's the best way. The Waterfall model isn't the only model that can be successful in the enterprise environment. Depending on the product and the expertise of your team, you might find that an Agile model or a hybrid blend of the Waterfall and Agile models would be successful.

If your team is right-sized — eight people or less — and you have multiple approval layers or end users, the Agile model might be the best fit. It works well when a small team works physically together in a location and has the ability to conduct face-to-face communication every day to minimize miscommunication and misunderstandings.

But keep in mind that if your team is small (again, eight or less) and you control the approval process, all other factors aside, you are positioned very well to implement RAD. In our opinion, RAD is the pinnacle of product development models. In an environment of unknown variables, RAD offers the most efficient, effective, and adaptable model for bringing a product to life.

Aside from the size of your team, here's our nutshell recommendation:

- Use Waterfall if you have a simple problem/product/idea that has a clear solution, requirements, and previous user feedback — combined with a larger team with lots of layers of approval built in and not a lot of flexibility.

- Use Agile if you have a more complex problem/product/idea that has some clear steps and requirements — based on user feedback — but you still don't fully understand what the solution is or will become.

- Use RAD if you have an even more complex problem/product/idea that doesn't necessarily have clear steps, requirements, or user feedback, and you don't know what the solution will become. This case applies to 99% of the innovators we interview, which is why we believe that RAD is the pattern to use for new and innovative thinking.

WATERFALL
RAD

Ultimately, choosing the right methodology and model is about finding the right tools for your team and product. The value and strength of every methodology resides in its structure. Regardless of your choice, own it: it's important to faithfully follow the standards and processes of that methodology. Like we said in the intro to this chapter — and it bears repeating here — don't switch product development models midstream unless you absolutely have to.

WITH THIS IN MIND, LET'S ACTUALLY GET INTO BUILDING YOUR PRODUCT.

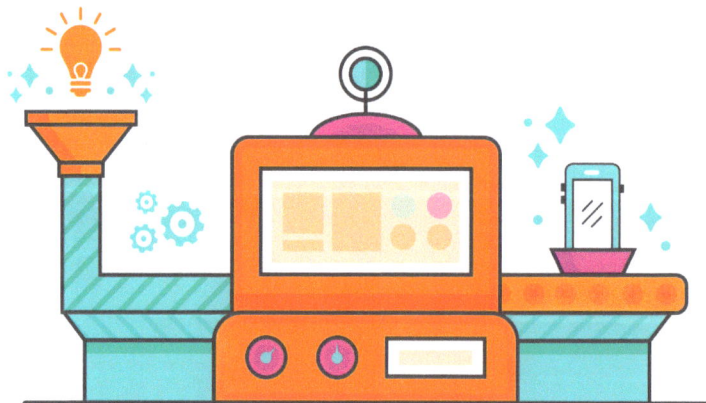

HOW TO INTERVIEW YOURSELF AROUND YOUR DEVELOPMENT METHODOLOGY

HOW MANY PEOPLE ARE ON YOUR TEAM?

DO YOU HAVE A PROJECT MANAGEMENT OFFICE (PMO)?

ARE YOUR DEVELOPERS ALREADY FAMILIAR WITH A MODEL?

DO YOU HAVE A QUALITY ASSURANCE PROCESS?

WHAT STAGE ARE YOU IN WITH THE PRODUCT (NEW OR MAINTENANCE)?

DO YOU HAVE END USERS?

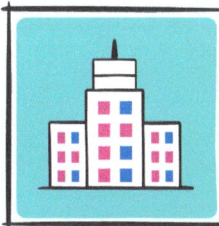

IS YOUR TEAM CENTRALLY LOCATED OR ARE YOU SPREAD ACROSS BUILDINGS?

DO YOU HAVE A SERVICE- LEVEL AGREEMENT (SLA) WITH END USERS FOR YOUR PRODUCT?

DO YOU HAVE A SINGLE PRODUCT OWNER WHO IS ACCESSIBLE AND ABLE TO MAKE QUICK DECISIONS?

HOW MANY LAYERS OF APPROVAL DO YOU HAVE?

"ANYONE WHO HAS NEVER MADE A MISTAKE HAS NEVER TRIED ANYTHING NEW."

ALBERT EINSTEIN

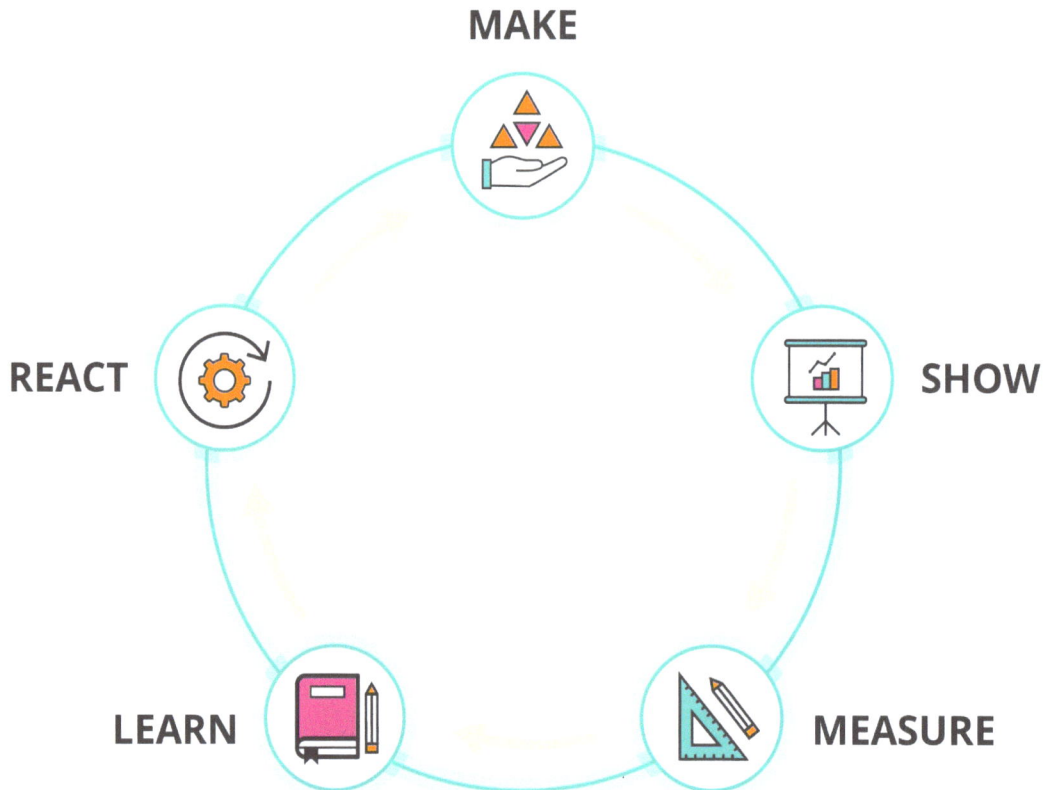

Building a product is an art. Art is an expression of our imagination. We don't want to detract from this magic; we want to harness and direct our creative energy.

But (big but) it's not all about pure creativity. If your end goal is to express your imagination without any restrictions, developing products probably isn't the right arena for you. But if you want to build a product with a high chance of success, welcome to the iterative build process.

We are all building our ideas on the shoulders of giants. Your ideas are simply iterations on ideas that came before. And your ideas, in turn, will be iterated — both by you, as you develop your product, and perhaps beyond you.

There are many, many different ways to build products, and we covered three of the main ones in the last chapter. In this chapter, we'll focus on how we typically move through a project using the RAD methodology, an iterative process where it's possible to change course in small and drastic ways throughout the build.

The British software developer and public speaker Martin Fowler revolutionized software development with the concept of failing fast. This is an idea that applies well to product development. But how to fail best?

In the iterative build process, a failure is measured the same way as a success. They are both conditions that require point decisions, like whether to continue or change course. Your main objective during product iteration is to move quickly so that you are in the "decision condition" frequently. At Google, they have a mantra: "Don't tell me. Show me."[34] In the early days of trying to grow a user base quickly, Facebook developers preached the motto "Move fast and break things."[35] At the core of the RAD build process is speed.

WITH AN ITERATIVE BUILD PROCESS, YOU START SMALL, GET FEEDBACK OFTEN, AND ADJUST ACCORDINGLY.

You'll start with a proof of concept, move on to the stage where you have a minimal viable product, then eventually get to your alpha, beta, and release candidates (see box). During each of these stages of building a product, there are five actions you'll take over and over and over:

1. MAKE
2. SHOW
3. MEASURE
4. LEARN
5. REACT

34 *How Google Works*, Eric Schmidt and Jonathan Rosenberg, Grand Central Publishing, 2014.
35 Samantha Murphy, "Facebook Changes Its 'Move Fast and Break Things' Motto," Mashable, April 30, 2014, https://mashable.com/2014/04/30/facebooks-new-mantra-move-fast-with-stability/#NPchK8_o9Pq

The cyclical nature of these activities is where innovation comes from. Conducting each action during each prototype phase will help save you time and money. It will ensure you are testing ideas as you go and, ultimately, innovating better products that people want to use.

> ## "TAKE RISKS. ASK BIG QUESTIONS. DON'T BE AFRAID TO MAKE MISTAKES; IF YOU DON'T MAKE MISTAKES, YOU'RE NOT REACHING FAR ENOUGH."
> ### DAVID PACKARD

These are all prototype phases that help save time and money by making sure you're testing ideas as you go and building the right things. Ultimately, this is where innovative products come from.

Google is the classic example of a company founded on and devoted to the spirit of innovation. The company's one-of-a-kind research lab, Google X, funds wild ideas like the driverless car and Google Glass. Innovation is at the core of the spirit that makes the company so successful. But Google can afford to work on wild ideas ad infinitum, because the company's search ad revenue is enormous and subsidizes everything else they do. By one account, in 2016, "Google's ad business accounted for 89% of Alphabet's revenue, or $76.1 billion."[36] If you aren't Google, you probably don't have such deep pockets to endlessly pursue ideas that may end up going nowhere. If that's the case, the iterative build process isn't just a conceptual philosophy; it's a financial mandate.

Building finished products can be tremendously challenging and can cost a mini-fortune. On the other hand, building small, iterative prototypes can be easy and cheap. The 80/20 rule applies here. You can build 80 percent of your features with 20 percent of the effort it will take to build the entire thing. The process of completing the product is where you will spend the most time and effort.

It's important to note that there will be many iterations within each phase. During each iteration cycle, you'll notice five distinct steps: Make, Show, Measure, Learn, and React. The rest of this chapter will cover these five distinct steps in detail.

---

36 Max Chafkin and Mark Bergen, "Google Makes So Much Money, It Never Had to Worry About Financial Discipline — Until Now," *Bloomberg Businessweek*, December 8, 2016, https://www.bloomberg.com/news/features/2016-12-08/google-makes-so-much-money-it-never-had-to-worry-about-financial-discipline

MAKE

REACT

SHOW

PROOF OF CONCEPT
MINIMAL VIABLE PRODUCT
ALPHA
BETA
RELEASE CANDIDATE

LEARN

MEASURE

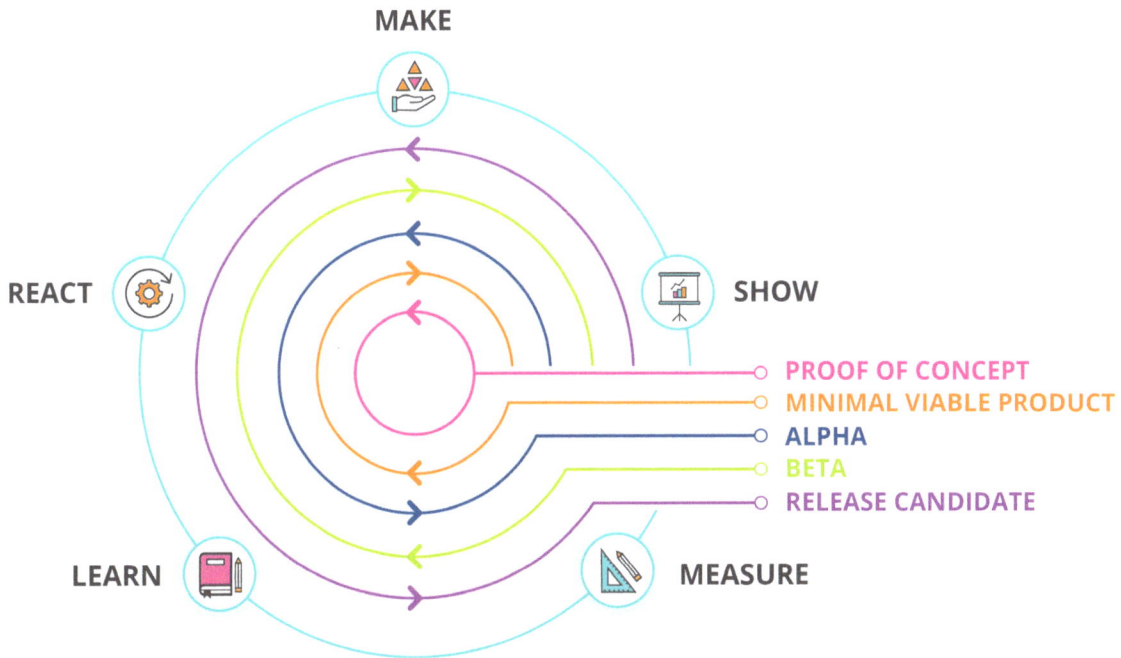

As you probably gathered from the last chapter, we're big fans of the RAD methodology of product building. Within that construct, we typically build in a cyclical way that involves five main stages of creation. Each stage can have multiple iterations within it. Projects can differ, of course, but in general, here's what the process tends to look like:

1. PROOF OF CONCEPT

Depending on what stage of iteration you're in, this might be anything from a sketch on a napkin to a deck on a tablet. It's a prototype you can use to prove your idea and test its rudimentary features within conversation with a "user."

2. MINIMAL VIABLE PRODUCT

Even at this early stage, your product should have a laser focus on user experience. If you're building an app, for instance, it doesn't need to be fully functioning and correctly coded at this point. In fact, it could still just be a series of screens that you walk the test subject through to "fake" the real experience. You just need a prototype that you can begin to test in front of users, cheaply and speedily.

3. ALPHA

By the time you get to the Alpha stage, your minimal viable product should be thoroughly coded on the platform and in the language you plan to release it on. It should also be testable by a small group of users — an active group of up to ten people who can give you concrete feedback. But it's still too unstable to send out to the general public.

4. BETA

In the beta phase, your focus is on squashing bugs. Features should be locked down; this isn't a time to add new things. Beta is about refinement. You might want to open this version up to the public, with a strong caveat that it's a work in progress. It's not ready for prime time yet. We recommend making this an invite-only release. Pick your users carefully and implement easy feedback mechanisms.

5. RELEASE CANDIDATE

This is your launch-ready product. It doesn't have to contain every feature you've ever dreamed of, but it does need to work, and to solve a specific problem for a user. It should be stable and relatively free of bugs, at this point.

Arguably, there is another stage called The Final Product. This is a fully functional, bug-free product that is ready for market. But even once a product has gone to market, the iteration never stops.

WITHIN THIS NEW ITERATIVE BUILD PROCESS, EACH OF THESE STAGES INVOLVES MAKE/SHOW/MEASURE/LEARN/REACT ACTIVITY.

PRINCIPLES OF ITERATION

1. BUILD PROTOTYPE AFTER PROTOTYPE, UNTIL YOU HAVE ENOUGH KNOWLEDGE TO BUILD SOMETHING GREAT.

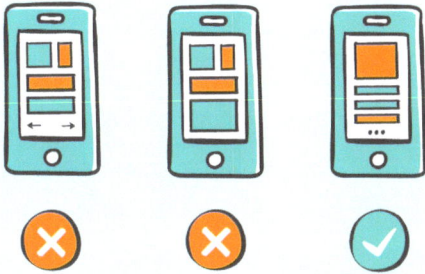

4. CONSIDER FAILURE A STATE OF LEARNING. EMBRACE IT. CHAMPION IT AS A VALUABLE SOURCE OF NEW KNOWLEDGE. AND ALWAYS BE WILLING TO FAIL FAST— AS FAST AS YOU POSSIBLY CAN WHILE STILL LEARNING FROM EACH FAILURE.

MAKE, SHOW, MEASURE, LEARN, REACT. **2.**

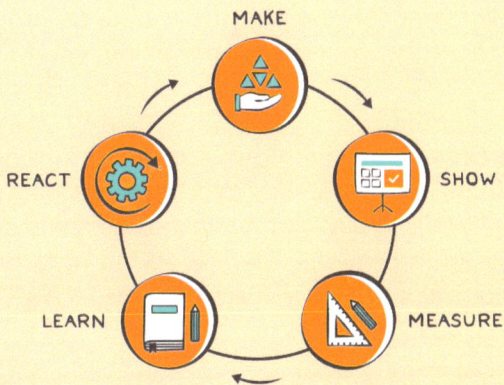

MAKE

REACT

SHOW

LEARN

MEASURE

5. VALUE EACH PERSON'S INPUT. ESPECIALLY ON A LEAN TEAM, EVERY SINGLE PERSON CONTRIBUTES BEYOND THEIR BASIC SKILL SET. THE MORE INPUT GATHERED, THE HIGHER THE OUTPUT.

3. VALUE MESSY EXPERIMENTATION OVER TIDY DESIGN— A MESSY BUT PARTIALLY FUNCTIONAL PROTOTYPE IS WORTH FAR MORE THAN A COMPLEX, POLISHED DESIGN THAT ONLY EXISTS ON PAPER.

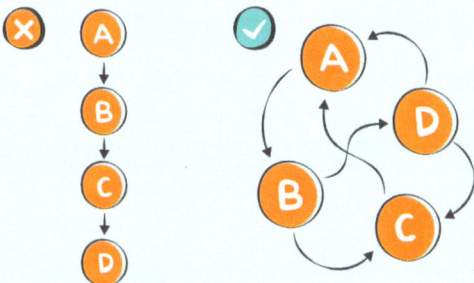

A
B
C
D

A
D
B
C

6. NEVER UNDERESTIMATE AN ABILITY TO PIVOT SWIFTLY. ALWAYS BE WILLING TO THROW AWAY IDEAS, LISTEN TO CONTRARY OPINIONS, AND MOVE QUICKLY. IT'S NOT ABOUT YOUR EGO; IT'S ABOUT WHAT'S WORKING.

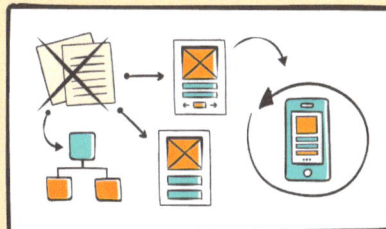

The first action you'll engage in within the iterative cycle is to build, or make. As the holder of the vision for the product, this is how you bring the product to life. The Make activity should have clear features that are testable in later iterations.

Your product vision may be grandiose, but at this point, it most likely lacks data to inform you if it's a worthy idea or not. To get the data to inform your decision-making, you must build the minimum features and iterate on them.

It's best to start simple. After all, the iterative process is all about gathering data quickly. It's sort of like playing poker: before you graduate to high-stakes gambling, it's best to play a bunch of rounds at low stakes. That way, you'll learn from your mistakes without it costing too much, until you're ready to make an informed, experienced bet at higher stakes. There's no substitute for experience, and that's what a low-stakes approach gives you.

DURING THE PROOF-OF-CONCEPT PHASE YOU MIGHT SIMPLY USE A PEN AND A NAPKIN TO SKETCH OUT YOUR IDEA.

THAT'S ALL YOU NEED TO START TO ITERATE.

Initial Proof of Concept Example

> "WHEN WE CREATE STUFF, WE DO IT BECAUSE WE LISTEN TO THE CUSTOMER, GET THEIR INPUTS, AND ALSO THROW IN WHAT WE'D LIKE TO SEE, TOO. WE COOK UP NEW PRODUCTS. YOU NEVER REALLY KNOW IF PEOPLE WILL LOVE THEM AS MUCH AS YOU DO." — STEVE JOBS

Let's define what the action of Make looks like in different phases of product development:

1. **Proof of Concept (PoC):** In this phase, Make may look like an idea on a napkin.

2. **Minimal Viable Product (MVP):** In this phase, Make is a functional prototype with very basic functionality.

3. **Alpha:** In this phase, Make is a buggy, unstable product. It has limited features, built for a closed-testing team.

4. **Beta:** In this phase, Make may still contain bugs, but they are unknown. It is a full-featured product, built for customer testing.

5. **Release Candidate:** In this phase, Make is nearly a final product; it just needs a final round of user testing.

QUESTIONS TO ASK YOURSELF

During the action phase of Make there are three questions you can ask yourself about your product and users:

1. What do we need to add to the user experience?
2. Where will they use it?
3. What can we fake with prototyping?

Once you get to the MVP phase, these questions morph. At that point, in addition to the questions above, you need to ask yourself:

1. Do the changes I've made actually solve the needs of the user and address what I've learned so far?

The next action, during each phase, is Show.

4.2 SHOW

The second action within the iterative cycle is Show. The main objective of Show is to gather honest, unrestrained feedback. You have to set the stage for this to happen effectively.

First, it's crucial to remember that you are looking for feedback on your product — it's not about you. To embrace negative feedback, you must separate yourself from the product. Colin Powell has some advice on this: "Avoid having your ego so close to your position that when your position falls, your ego goes with it."[37]

Second, you need to show your product to the right group of people during each development phase of the project. Find trusted groups you can brag about your product to. Depending on the phase you're in, such a group might be comprised of peers, investors, or prospective end users. During all of the phases of development, it's important to not argue, defend, dismiss, or minimize the feedback from anyone willing to give it. Take the feedback and thank whomever provided it.

Let's see what Show might look like in the different phases of product development:

1. **Proof of Concept (PoC):** Remember that idea you jotted down on a napkin? In this phase, Show means letting someone else take a look at it.

2. **Minimal Viable Product (MVP):** In this phase, you have a crude product with just enough functionality and features to convey the spirit of the idea (Show it) to a trusted group of advisors and investors.

3. **Alpha:** Now you have a product to Show a closed group of users. The product is missing features, and it has bugs, but you know this. You know it's flawed.

---

37 Colin Powell, *It Worked for Me: In Life and Leadership* (Harper Collins, 2012).

4. **Beta:** You have a full-featured product to Show off. There are probably still bugs, but you don't know what they are. Your audience might be a closed group of users or open user testing.

5. **Release Candidate:** Finally, you have a full-featured, bug-free product that has been user tested. You're ready to Show it to end users for what you hope will be the last round of user testing.

At some point, as your product grows and you find yourself with a few hundred active users, it will cease to make sense to solicit feedback from every single one. You will become quickly overwhelmed with user feedback if you do. And users, as we've mentioned, aren't always adept at expressing succinct, useful feedback. So somewhere around the MVP or Alpha stage, you'll have to start applying your "vision filter" to every bit of feedback you solicit. Otherwise, you'll try to accommodate every whimsical request and quickly end up with a worse problem: feature bloat.

QUESTIONS TO ASK YOUR "USERS"

During the show activity, ask your users three questions:

1. Do you know how to _____?
2. Is it easy to ____?
3. How can we make the experience better?

As you move through the phases and tie everything together — confident you have a product-market fit — you might also ask:

1. Is my product still as simple and focused as it could be? (Go back to the Principles of Human-Centric Design in 1.7.)
2. Would you share this with a friend?

The next action is Measure.

4.3 MEASURE

The third action in the iterative cycle is Measure — gathering data from users along the way in order to improve and troubleshoot. The Measure activity requires you to be unbiased and objective. There isn't a right or a wrong. This activity is purely concerned with gathering relevant data points. The meaning of relevant data points is different based on the phase you are in.

"IF YOU CAN NOT MEASURE IT, YOU CAN NOT IMPROVE IT."
LORD KELVIN, A.K.A. THE PHYSICIST WILLIAM THOMPSON[38]

1. **Proof of Concept (PoC):** In this phase, you Measure whether your product concept is feasible. Relevant data points refers to feedback from your trusted group.

2. **Minimum Viable Product (MVP):** Here, you Measure the viability of your product. You aren't concerned with asking anyone outside of your trusted group of advisors and investors.

3. **Alpha:** At this point, you Measure limited product features and functionality. Measure looks like user feedback — e.g., bug and usability reports. Relevant data points are those from your closed group of users.

---

38 "William Thomson, Baron Kelvin," *Encyclopedia Britannica*, accessed July 20, 2017, https://www.britannica.com/biography/William-Thomson-Baron-Kelvin

4. **Beta:** In this phase, you Measure all of the features and functionality of your product with a group of users, gathering relevant data points only from this group. Measure looks like bug and usability reports.

5. **Release Candidate:** Measure is similar to the Alpha and Beta phases. The main difference is that in this phase, there is a very broad group of users to gather data from — all users, in fact. Relevant data points are open to everyone using the product.

QUESTIONS TO ASK YOURSELF

During the Measure activity, ask yourself three questions:

1. Are we measuring the right key performance indicators (KPIs)?
2. Are we gathering only the minimum critical data?
3. Are we missing something?

As you progress through product cycles, for instance, here are some key engagement metrics you might measure:

- Active users (daily and monthly)
- Monthly growth (new and active users)
- Bounce rate
- Conversion rate
- Churn

Measurement is crucial to improvement, but the Measure activity alone does not create positive change. For that, you need to move on to the next activity: Learn.

4.4 LEARN

The fourth action within the iterative cycle is called Learn. We've taken in all the data from the Measure activity, and gained an understanding about the product, at whatever phase it's at. Now, it's time to study that data in order to gain knowledge.

> "HE WHO NEVER DOES ANYTHING WRONG, SELDOM, IF EVER, DOES ANYTHING RIGHT." — LEO TOLSTOY

During this activity, it's important to always remember that all knowledge is good. The more knowledge, the better. It's so common for the "Learn" action to be a challenging one for entrepreneurs. We have fixed ideas about our products that are held aloft by our ego. Users might be telling us something pretty loudly, but we refuse to hear it because it conflicts with our ideas about our self, our mission, and our product.

You must be inquisitive, neutral, and willing to learn. Knowledge helps you make decisions, and making decisions makes for better products. The whole point of these actions is to get you to a decision condition as often as possible throughout the lifecycle of your product design.

1. **Proof of Concept (PoC):** In this phase, Learn means relying on the data you gathered to figure out if you have a valid concept and whether it will be feasible to build a prototype. This is a "make or break" stage — literally. You're either going to make the product, or you're going to break the project.

2. **Minimum Viable Product (MVP):** Here, Learn means to further understand the nuances of your product's viability. At this point, you should have a solid idea how your investors and advisors feel about your product.

3. **Alpha:** At this point, Learn comes from the knowledge you gain with your close user group's feedback. You should have lucid ideas around usability and bugs.

4. **Beta:** In this phase, Learn looks much like it does in the Alpha phase. The difference is that you have now collected data from a larger group of users, and have an understanding of any outstanding bugs and usability issues.

5. **Release Candidate:** In this phase, Learn gives you further understanding about your product. All bugs should have been addressed, so the main focus of your learnings is on product fit and usability.

QUESTIONS TO ASK YOURSELF

During the Learn activity, ask yourself three questions:

1. What interactions are working for the user?
2. What user flows are not working?
3. Is there a new or better idea we can test?

As you hone your product, also ask yourself:

1. Is my product still closely aligned with my vision? (Go back to Translating Vision into Design in 1.6.)

And finally, let's talk about React.

4.5 REACT

With this activity, we React to everything we've learned in order to answer the question "Should we stick, pivot, or quit?"

If you're getting positive feedback about your product, and it's solving a real problem people have, stick with it.

If your product is solving a real problem, but you aren't getting traction with user adoption, then you might just have a marketing, UX, content, or messaging problem to solve. You are still on the right track, so continue to move forward into the next iteration. Again, stick.

If you're at a point where you've been unsuccessful in raising funds, or you've already burned through all your capital, or your product simply isn't connecting with users, you're at a crossroads. Go back to your value proposition and apply the knowledge you've gathered from feedback since it was written. If you can't align the product's value proposition with the feedback, it's time to quit.

If you can, however, it's time to pivot.

Making this decision takes intelligence, courage, and fortitude. You have to take into consideration your product's value proposition and feasibility as well as your personal motivation and passion for the product — and any feedback.

Here's what React might look like at various stages of the iterative cycle:

1. **Proof of Concept (PoC):** In this phase, React means to take what you've measured and learned and use that information to modify your product idea. Incorporate the feedback you've gotten from your preliminary group of advisors and start the iterative process over.

2. **Minimum Viable Product (MVP):** Here, React means adjusting the alignment of your value proposition based on feedback. Once you've revisited your value proposition, assess whether it stays true to the product idea from the previous phase. If it does, continue to iterate. If it does not, and the product idea has changed dramatically, go back to the PoC phase. You'll know you're ready to move on to Alpha when you're confident you've found a product/market fit. (Incidentally, this is an ideal time to start talking to investors.)

3. **Alpha:** At this point, user feedback might cause you to discover that crucial features of your product aren't needed or technically feasible, so you'll need to realign the value proposition and move back to the MVP phase.

4. **Beta:** At this phase, a general pivot is unlikely because you've hopefully learned enough from the MVP and Alpha phases to be confident that your product is working, at least for now. Here, it's more likely you'll be making small feature and usability changes.

5. **Release Candidate:** User testing will inform the way you React. Again, a major pivot is unlikely and only happens when you suddenly discover you have a critical, unfixable bug or another deal-breaking usability issue.

One of the main reasons companies end up pivoting is because the problem they were trying to solve for users turns out to not actually be a problem at all. Sometimes, what we think is a problem isn't. Even if a product solves a problem well, if it's not something people are trying to fix, they're not going to go for the product.

Know when to stop wasting your own time and others' time. If it's time to quit, have the courage to do so. But if a pivot is a realistic option, you, as the vision holder of the product, must be driven by passion. It's hard to give your all to a product you aren't passionate about. If you don't have the motivation, then stop. If you're burned out, take some time to recharge and regain focus. And when you're ready, re-engage on a product idea you're passionate about.

Are You Solving a Problem That Actually Exists?

- Justin Jones -

One of the most common startup failures is trying to solve a problem that doesn't exist, or that people don't want solved. We often tell wannabe product makers that solving a problem they personally have is a start. But that can backfire, too.

In my experience as a digital product maker, I've made this mistake more than once. When my Grandpa passed away, everyone in my extended family looked for pictures of him they could post online in tribute. There weren't a lot of photos in the first place, and the ones that existed were all over the place — not in a common location. Wouldn't it be wonderful, I thought, if there was a memorial site where everyone could post their common pics?

With Scott, I set out to create a "Facebook for the Dead" (don't laugh, this was our actual tagline). It seemed like such an innovative idea at the time; it would solve a problem I knew many people had when their loved ones passed away. But the app never generated a lot of traffic, and I eventually realized why: because although a central repository of images of your dead loved one is helpful when you need it, it's not a problem people wake up with every day. It's typically only a problem for a short period of one's life, so the solution simply can't generate a lot of use.

This was a valuable lesson for me. The project got killed early on, and I took away some deep insight about trying to solve problems that people don't really need solved.

5.0 THE SECRETS OF EFFECTIVE TEAMS

We're up to Chapter 5 and we're just now talking about building a team, but make no mistake, your team is vital to your vision and your business model and everything else we've talked about so far. From our experience, the only real way to achieve success with a product is with a team, and the only kind of team to have is an effective one. A collaborative team is like a marriage, and just like a marriage, it's easy to overlook red flags during the honeymoon phase. If you don't have the right people on board, it can get ugly when the going gets tough. The wrong team can ultimately be the downfall of your product.

For this reason, your core team, in particular, is vital. Many startups fail not because of the product, but because of the team being unable to effectively unify their skills behind executing a vision. As you grow and expand and add team members, you will have more room for error, but you'll also have less accountability. Both of these things, ironically, lead to lower productivity. Which is why, although it's definitely hardest to build a product entirely by yourself, it's nearly as hard to build one as part of a massive organization. The easiest way to build a product, by far, is with a small team with a unified vision.

In fact, many of the world's best ideas originate from small, lean, startup teams made up of individuals with overlapping skills who possess a high level of trust amongst each other and are highly collaborative. Let's break this down.

SMALL TEAM SIZE

In the beginning, your team should be anywhere from two to eight members. Keeping your team this small might be a challenge because you are balancing collective intelligence against social interactions. In other words, you need members with enough diverse skills to deliver the product. That collective intelligence is greater than any one member's talents.

But as you add more and more talented folks to your team, you create a higher level of social interaction — required for the team to make decisions, but expensive in terms of time and energy. More team members means higher collective intelligence, but more social interactions brings more risk of time wasted. There's a sweet spot, and we've found it to be in the two-to-eight range.

OVERLAPPING SKILLS

Each member brings something unique to the team, but there is depth to skills and then there is breadth — where skills overlap between team members. For example, you might have a designer and a developer on your team. The designer's depth skill is in visual or experience design, and the developer's depth skill is in writing code. But both have experience in UI — that's their shared breadth skill.

Because of their shared skill, you actually have two UI people on your team. The developer can offer a lot during the design phase, and the designer can offer a lot during the development phase. This is often referred to as T-shaped skills: a team member has depth in a single field (the vertical bar of the T, where the horizontal bar is breadth of skills) — but also, and this is important, an ability to collaborate across disciplines with other team members who are experts in other things.

HIGH LEVEL OF TRUST

On a team, everything is built on trust. You can have all the ingredients of an effective team, but without trust, it won't work. Each member needs to rely on the others. Each member needs to critique and be critiqued by the others. When there is a high level of trust, critique can be given and taken bluntly without offense. Trust reduces the social interaction cost.

There are many ways to build trust, but our favorite is through play. Play together, become friends, learn who you are working with, and share who you are with everyone else on your team. Get together outside the workplace and learn the names of the people important in your team members' lives. You cannot have an effective team without trust, period.

COLLABORATION

"THE BUSINESS WE'RE IN IS MORE SOCIOLOGICAL THAN TECHNOLOGICAL, MORE DEPENDENT ON WORKERS' ABILITIES TO COMMUNICATE WITH EACH OTHER THAN THEIR ABILITIES TO COMMUNICATE WITH MACHINES."

TOM DEMARCO[39]

Effective teams have effective interactions. You can only take advantage of the team's collective intelligence if team members are working together. As individuals, we are limited. We only see situations from a single perspective, and the more complex the problem, the more our brains take cognitive shortcuts. Sometimes these shortcuts result in brilliant ideas, but more often, they result in mistakes. When a team effectively collaborates, it makes less mistakes and produces higher-quality work.

IN THIS CHAPTER, WE'LL TAKE A LOOK AT HOW TO HIRE, MANAGE, AND MAXIMIZE THE PRODUCTIVITY OF AN INTEGRATED SMALL TEAM.

CELEBRATE DIVERSITY

Building a good team is kind of like assembling a football team. If you find the Tom Brady of engineers, great, but you'll still need to round out your team with offensive linemen, running backs, and receivers. You need people who can score and people who can block a 300-pound bull on the opposing team. Even the world's best quarterback is useless if not surrounded by an equally talented team with a diverse set of skill sets.

For this reason, it doesn't always work to simply hire and train clones of your best guy/gal. The magic happens when you pair leaders with extraordinary vision with a supporting cast of like-minded all-stars who shine in their respective disciplines and are not afraid to express their individuality.

---

39 Tom DeMarco, *Peopleware: Productive Projects and Teams*, (Dorset House Publishing Company, 1999).

5.1 DEVELOP AN INTEGRATED TEAM

"WHEN WE SENSE ESPRIT DE CORPS, THAT PERCEPTION DOESN'T COME OUT OF THE BLUE; IT'S THE RESULT OF OUR INNATE ABILITY TO PROCESS THE HUNDREDS OF COMPLEX COMMUNICATION CUES THAT WE CONSTANTLY SEND AND RECEIVE."
HARVARD BUSINESS REVIEW[40]

With an integrated team, every single person is involved from the git-go. Team members understand the vision and don't need to be brought up to speed. There's no learning curve, and the success rate is higher than with a team that's not integrated. The integrated team executes faster, produces higher-quality product, keeps costs down, handles a schedule well, produces excellent time to market, and is generally more efficient.

SOUNDS GOOD, RIGHT? BET YOU'D LIKE A TEAM LIKE THAT.

You are probably hiring some combination of product developers, designers, marketing folks, salespeople, and, of course, leadership. These are a lot of different types of personalities, and now is the time to lay the foundation for an integrated team. There are several key elements of integrated teams that you can put in place right from the start of your hiring process. The very first element is that there is a clear leader. We're guessing that's going to be you. It's your responsibility to assemble a team with the appropriate qualifications to succeed — then manage them well.

HERE ARE SOME MORE KEY ELEMENTS OF INTEGRATED TEAMS:

- There is individual and mutual accountability (both are crucial).
- The purpose of the team is clear.
- There is open-ended discussion and active problem-solving that involves the whole team.
- Team performance is measured as the collective work and success of the product.
- Work is delegated based on skills, knowledge, and availability.
- Team members have overlapping skills and individual specialties.

---

40 Alex "Sandy" Pentland, "The New Science of Building Great Teams," *Harvard Business Review*, from the April 2012 issue, https://hbr.org/2012/04/the-new-science-of-building-great-teams

THE BLUEPRINT FOR
RAMPANT INNOVATION

INNOVATION COMES FROM THE TOP. BUT HIRING THE RIGHT PEOPLE ENSURES THAT A CULTURE OF INNOVATION IS RAMPANT THROUGHOUT YOUR ORGANIZATION AS IT GROWS FROM ONE TO TEN TO A HUNDRED TO A THOUSAND PEOPLE.

ALONG THE WAY, MAKE YOUR COMMITMENT TO INNOVATION THE PLATFORM FOR EVERY DECISION, EVERY ACTION, AND EVERY BEHAVIOR.

1. FOCUS ON THE USER, AND ALL ELSE WILL FOLLOW— INCLUDING THE MONEY.

4. SHIP SWIFTLY, THEN KEEP ITERATING. DON'T GET STUCK IN LONG SPRINTS.

2. SHIFT YOUR DEFINITION OF FAILURE AND USE THE TESTING PROCESS TO CONSTANTLY ITERATE.

5. ELIMINATE SILOS WHENEVER YOU CAN. ENGAGE YOUR DEVELOPERS IN THE MARKETING PROCESS. INVITE EVERYONE WITHIN YOUR COMPANY "WALLS" TO COLLABORATE.

3. FIND YOUR EMPLOYEES' PARTICULAR PASSIONS, THEN SET THEM LOOSE ON PROJECTS YOU KNOW THEY'LL THRIVE WITHIN.

6. DON'T JUST TRY TO BE BETTER. AIM TO BE RADICALLY BETTER — AT LEAST 3X BETTER, IF NOT 100.

7. WITH THE RIGHT VISION, GIVE EVERYONE AN OPPORTUNITY TO BECOME A PART OF SOMETHING MUCH BIGGER THAN THEMSELVES.

But there's also one more very important element of integrated teams we haven't mentioned yet: a willingness to fail. Even the most successful and wildly profitable products endured countless failures and missteps before success took shape. Your team must be willing to fail. Over and over again, if necessary. In fact, the experience of failing together will be the glue that binds your team.

> "WHEN MANY OF US THINK ABOUT INNOVATION, THOUGH, WE THINK ABOUT AN EINSTEIN HAVING AN 'AHA!' MOMENT. BUT WE ALL KNOW THAT'S A MYTH. INNOVATION IS NOT ABOUT SOLO GENIUS, IT'S ABOUT COLLECTIVE GENIUS."
> LINDA HILL IN HER TED TALK "HOW TO MANAGE FOR COLLECTIVE CREATIVITY"[41]

Getting up after being knocked down is what ultimately allows teams to deliver optimal products and overcome impossible-seeming obstacles. If you avoid failure, and mitigate your risk too carefully, you limit your own potential.

TO BE MORE SUCCINCT: WITHOUT FAILURE, YOU CAN'T HAVE SUCCESS.

On the other hand, you don't want to fail too much. Studies have shown that projects tend to fail because of human, rather than technological, factors. Putting an integrated team and practices into place up front helps ward off ultimate failure.

---

41 "Linda Hill: How to manage for collective creativity," *TED*, accessed July 7, 2017,
    http://www.ted.com/talks/linda_hill_how_to_manage_for_collective_creativity/transcript?language=en

"IT IS HARD TO FAIL, BUT IT IS WORSE NEVER TO HAVE TRIED TO SUCCEED."
TEDDY ROOSEVELT

"IT IS HARD TO FAIL, BUT IT IS WORSE NEVER TO HAVE TRIED TO SUCCEED."
TEDDY ROOSEVELT

TO LEAD A PRODUCT, COMPANY, OR TEAM, YOU NEED TO BE A PASSIONATE LEADER WITH A VISION FOR THE FUTURE.

This enables you to attract the talented people who you will need to help bring your vision to life. And then you need to remove any barriers that stand in their way (human, technology, etc.).

Kim Scott, CEO of Candor, Inc., and renowned coach for companies the likes of Twitter and Qualtrics, talks about radical candor in guiding employees — a concept that applies not just to how bosses treat their teams, but to how peers treat each other within an organization.

Radical candor is about offering guidance in the form of both praise and criticism — both are crucial. But in order to get to the point where employees can hear them equally and openly, you must first develop relationships.

THE FIRST STEP to creating a positive relationship with an employee is to demonstrate that you care about them. This might sound like a touchy-feely bit of fluff, but in fact it's a crucial and pragmatic step. Get to know your employees. Find out if they have families. Know what's important to them. And help them focus on their priorities — both within and beyond the office.

THE SECOND STEP to creating a culture of "radical candor" is to be willing to challenge people directly. As Kim Scott reminds us, we've all been taught since Kindergarten "if you don't have anything nice to say, don't say anything." Well, this is a nice idea, but in business, it's simply not true. In business, it's not just your job to criticize; it's your moral obligation.

SCOTT DESCRIBES RADICAL CANDOR AS THE INTERSECTION OF CARING PERSONALLY AND CHALLENGING DIRECTLY.[42]

THE OBLIGATION TO DISSENT

"YOU CAN'T BE AN EFFECTIVE LEADER IN BUSINESS, POLITICS, OR SOCIETY UNLESS YOU ENCOURAGE THOSE AROUND YOU TO SPEAK THEIR MINDS, TO BRING ATTENTION TO HYPOCRISY AND MISBEHAVIOR, AND TO BE AS DIRECT AND STRONG-WILLED IN THEIR EVALUATIONS OF YOU AS YOU ARE IN YOUR STRATEGIES AND PLANS FOR THEM."

BILL TAYLOR, HARVARD BUSINESS REVIEW[43]

McKinsey & Company is a powerhouse global consulting firm with over 12,000 consultants and 2,000 research and information professionals. The company is owned by 1,400 partners and is sans a central headquarters.[44] Truly a modern company devoted to collaborative innovation, McKinsey has a very progressive approach to HR. And one of the core values in their company mission is a dedication to "uphold the obligation to dissent."[45]

---

42 "Radical Candor — The Surprising Secret to Being a Good Boss," *First Round Review*, April 4, 2016, https://docs.google.com/document/d/1saJiLX2FMk5ZMfKsIe3oNB6lN8a3zbEEpsvepTng1ac/edit
43 Bill Taylor, "True Leaders Believe Dissent Is an Obligation," *Harvard Business Review*, January 12, 2017.
44 "Who We Are," McKinsey.com, accessed July 7, 2017, http://www.mckinsey.com/about-us/who-we-are
45 "Our Mission and Values," McKinsey.com, accessed July 7, 2017, http://www.mckinsey.com/about-us/what-we-do/our-mission-and-values

You have an ethical mandate to encourage those around you to speak their minds, even when they disagree with you, with each other, and with your mission. Every person in every room has a valuable point of view, and if your most junior person or your newest hire doesn't feel comfortable speaking up, you're missing an enormous opportunity for valuable input.

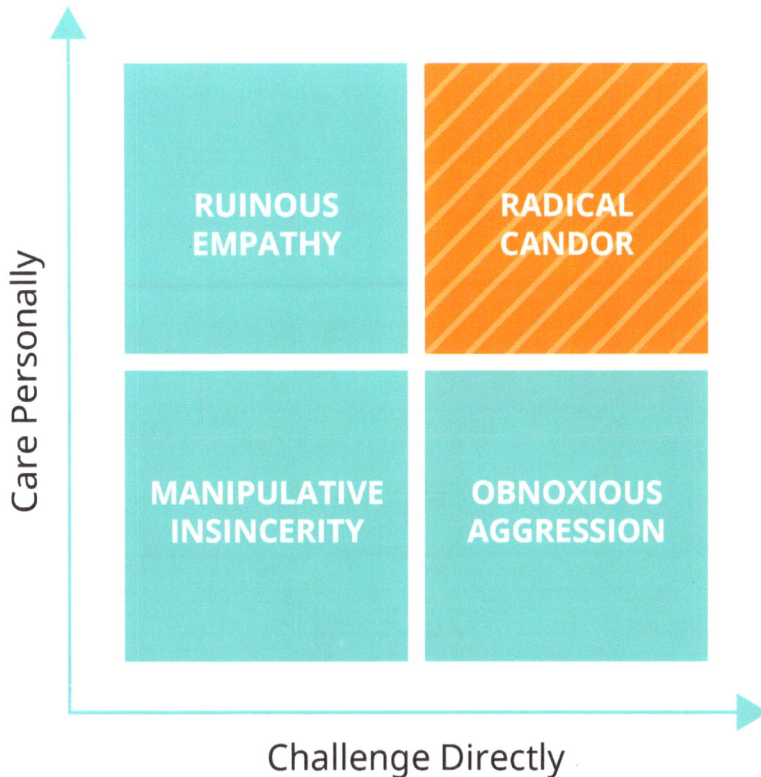

Here are some pointers around criticism that keep it constructive.

✓ Criticism should be:

○ Courageous and outspoken

○ Humble yet helpful

○ Impromptu and immediate

○ From a place of caring

○ Private if negative

○ Public if positive

It's important to note that radical candor cannot be a top-down approach. Bosses, employees, contractors, investors — all of these people must be willing to take an equal part in a culture of constructive criticism. No matter what your position or status in a company — and no matter the size and status of the company — you can model this behavior starting right now.

MORE ON RADICAL CANDOR

If you'd like to read more about radical candor and Kim Scott, we recommend Scott's book *Radical Candor*, the First Round article "Radical Candor — The Surprising Secret to Being a Good Boss," and the YouTube video of the same name.

5.3 STAY SMALL, BUT DELIVER BIG

"ADDING MANPOWER TO A LATE SOFTWARE PROJECT MAKES IT LATER."
FRED BROOKS[46]

---

46 Fred Brooks, *The Mythical Man-Month: Essays on Software Engineering* (Addison-Wesley, 1975).

If you feel stuck with a small team, don't. Small teams are actually more productive than big teams — at least at the product development stage. You could almost say that a smaller team increases your odds of a successful product.

As an environment grows bigger and becomes corporatized, an interesting thing happens: management begins to perceive themselves as facilitating a process, yet the creators suddenly encounter more and more roadblocks.

Jeff Bezos, the CEO of Amazon, famously quipped that if a team can't be fed with just two pizzas, the team is too big.[47] It's a common beginner's error in the software world to believe that throwing more manpower at a project will make it go faster and turn out better; the opposite is often true. This is because as teams grow, more issues arise.

> "AS ORGANIZATIONS STRIVE TO STAY AGILE AND INNOVATIVE, THEY'VE DISCOVERED THAT UNITS OF 8 TO 12 PEOPLE WORK BEST AS THE NATURAL SIZE OF HIGH-PERFORMANCE TEAMS. THIS IS THE MAGIC NUMBER FOR LEADERSHIP TEAMS, PRODUCT TEAMS, RESEARCH TEAMS, DESIGN TEAMS, AND MORE."
>
> ## RICH KARLGAARD[48]

Social psychologist Bibb Latané famously studied the phenomenon known as social loafing. As groups get larger, individual members experience less pressure to perform — and less perception of mattering — which leads to active disengagement.[49] At the same time, individuals in growing teams experience something called relational loss: the perception of receiving less and less support. This leads to feelings of isolation and chronic stress, which ultimately leads to poorer performance.

When teams get too big, employees begin to disengage, progress gets stalled, and, ultimately, the culture is ruined. Unanimously, people begin to do less — often, just enough to get by — while they start thinking about places they can go to advance their career with the "next big move." In short, the dynamic startup mentality is gone.

SO: SMALL TEAMS. BETTER.

---

47 Janet Choi, "The Science Behind Why Small Teams Work More Productively: Jeff Bezos' 2 Pizza Rule," *Buffer Social*, July 29, 2013,
    https://blog.bufferapp.com/small-teams-why-startups-often-win-against-google-and-facebook-the-science-behind-why-smaller-teams-get-more-done
48 Rich Karlgaard, "8 Reasons Small Teams Work Better," *Govexec.com*, April 18, 2014,
    http://www.govexec.com/excellence/promising-practices/2014/04/8-reasons-small-teams-work-better/82804/
49 Abstract of article, *APA PsycNET*, accessed July 7, 2017, http://psycnet.apa.org/journals/psp/37/6/822/

SMALL TEAM DYNAMICS

Of course, with small teams come particular team dynamics. In addition to creating a culture of radical candor, you have to work with those dynamics and facilitate collaboration. Here are some of the things that will help you better manage a small team.

1. A light-touch management style
The least amount of management you can possibly get by with is ideal.
Managers shouldn't make people work. They should make it possible for people to work.

2. The right tools
Cloud-based tools that allow people to communicate, share, and collaborate are ideal, particularly for workers who may not be reporting to desks in an office in a single location every day. (See box for some of our current recommendations.)

3. The right cadence to meetings
Too many meetings means no real work is getting done. Interruptions kill productivity, so make sure your creators have plenty of time every day that isn't broken up by unnecessary meetings. (Go look for the YouTube video "One Googler's Take on Managing Your Time" about how managers and makers schedule their days differently.) On a similar note, you don't need meeting minutes. Once the meeting is over, take a picture of the whiteboard, upload it to your cloud server, and you're done.

4. Opportunities to get to know people better
Social events, or just inline chances to chat at work, are important. But tackling challenges together outside of work — things like running a race or competing as a team in a scavenger hunt — can up the ante on group bonding.

5. A practice of transparency
Make knowledge accessible, motives obvious, and decisions visible.

6. Frequent feedback
Everyone on your team is participating in both talking and listening, and most of these conversations happen face to face.

7. Absence of politics between team members
Internal politics increase anxiety and effort, and tend to stop productivity in its tracks — and they are way more common in bigger companies. This, by the way, is one of the main reasons that small, underfunded startups are able to compete with large, resource-rich organizations. Self-created roadblocks between team members kill morale and mean much less gets done.

More than anything else, teams that "click" know how to communicate. MIT's Human Dynamics Laboratory studied group dynamics in professional settings to figure out just what differentiates a high-performance team. They found that the best predictors of productivity were the way a team interacted and engaged outside of formal meetings. They then deduced that small adjustments to the workplace culture — things like giving all employees a break at the same time — encouraged them to get to know each other and socialize away from their computers or workstations.[50]

When one call center tried this tactic, they predicted $15 million a year in productivity increases and also saw employee satisfaction rise more than 10%.[51]

Within a small team, you can have all the expertise you need. But team cohesion is always more important than experience and knowledge. Keep your team small for as long as you can — and capitalize on the power of a streamlined team.

OUR PERSONAL FAVORITE TRIFECTA OF TOOLS

In the digital product space, creating the right environment for productivity and collaboration is crucial. People who come to product making from a traditional business environment often overlook this, but it's essential to maximizing the productivity and communication of your people.

One of our favorite tools is Trello. The creators of Trello didn't start out saying "We want to make a visual collaboration tool and sell it." They started instead with a vision of building a place where talented developers would want to work and providing the tools to enhance the experience. They built out a private work space and gave each developer two monitors. One of the better products that came out of this effort was Trello. We use Trello for project management and we love it.

We also use Slack for communication. Both of these tools have desktop and mobile apps so the team can be connected anywhere/anytime. They also integrate with each other, so anything happening on the Trello board can be posted to the Slack channel. It's an efficient, cheap, easy setup — there's the right balance or organization and ease of use.

Rounding out our tool trifecta, we use GitHub for code.

---

50 Alex "Sandy" Pentland, "The New Science of Building Great Teams," *Harvard Business Review*, from the April 2012 issue, https://hbr.org/2012/04/the-new-science-of-building-great-teams
51 Ibid

"IF YOU HIRE PEOPLE JUST BECAUSE THEY CAN DO A JOB, THEY'LL WORK FOR YOUR MONEY. BUT IF YOU HIRE PEOPLE WHO BELIEVE WHAT YOU BELIEVE, THEY'LL WORK FOR YOU WITH BLOOD AND SWEAT AND TEARS."

SIMON SINEK[52]

During a recent visit to Facebook's headquarters, we heard a member of the company's executive team talk about the importance of "the Facebook mission" and how it's critical to the company's ability to attract and recruit talented engineers. This is likely one of the reasons Facebook updated its mission in June 2017: "Give people the power to build community and bring the world closer together."[53]

Facebook's original mission, "Making the world more open and connected," "had one fundamental flaw," according to Josh Constine at TechCrunch, "It didn't push for any specific positive outcome from more connection."[54] That minor mission statement tweak infused the company's hiring intentions with more defined purpose.

Facebook CEO Mark Zuckerberg preaches a simple litmus test for hiring: "Never hire someone to work for you unless you would work for them." To practice this philosophy takes a level of self-awareness and confidence. We know lots of people who talk about hiring people smarter and better than them, but inevitably, end up hiring others who are clearly subordinate and many times their junior in experience and skill.

---

52 "How Great Leaders Inspire Action," *TED*, accessed July 20, 2017,
    https://www.ted.com/talks/simon_sinek_how_great_leaders_inspire_action/transcrip
53 Mission Statement Page, *Facebook*, accessed July 20, 2017, https://www.facebook.com/pg/facebook/about/
54 Josh Constine, "Facebook changes mission statement to 'bring the world closer together'," *TechCrunch*, June 22, 2017, https://techcrunch.com/2017/06/22/bring-the-world-closer-together/

Why does this happen? Hiring someone smarter than you is threatening. You have to have the self-confidence to hire someone who will challenge you and your ideas. People who are talented can often be forces of nature — and difficult to manage. But as you're building a new product, it's essential to have a mental model for talent acquisition that will lead to a high-performance team. You're hiring toward a unified vision, but you're not assembling an army of mindless robot clones. Just the opposite, in fact. Curate a diverse group of individuals who believe in your purpose, which should in turn be an extension of your own personal belief system, but hire people who are independent thinkers.

So why does Facebook put so much effort into attracting the right people? Because like any company, but particularly a company on the cutting edge of innovation, people are the most valuable asset. As a startup, you might be leasing an office space, renting office furniture, and keeping all your intellectual property (IP) in the cloud — if you even have an office. Truly, your only commodity is your human capital. It's always best to double down on hiring the right talent, even if it takes longer or costs more. In the long run, it won't. There's synergy when you surround yourself with an A-team of players. Excellent talent lifts everyone up; underperformers, while potentially easier to manage, drag everyone down.

As any HR specialist will tell you, hiring right in the first place is the way to retain talent and increase employee satisfaction and effectiveness. The competition for tech jobs is fierce, and money is less of a factor in hiring than ever before. Candidates want to be a part of something bigger in what Aaron Hurst has coined "The Purpose Economy."[55] An iteration on the Age of Information, the Purpose Economy defines the professional motivations of millennials. And for most millennials, that purpose has something to do with making a positive impact on the world. The people who make up today's tech workforce strive to be a part of something bigger and more important than their individual job function.

This is a major shift in the mindset of many organizations. Throughout the Industrial Age, operating processes were always geared toward minimizing risk and eliminating mistakes. This makes sense in a factory or manufacturing setting, of course. But in the digital era, not so much. A culture designed to eliminate risk and banish mistakes stifles creativity and innovation.

TODAY, THE DEFINING CHARACTERISTIC OF MOST SUCCESSFUL COMPANIES IS AN ABILITY TO CONTINUALLY DELIVER INNOVATIVE PRODUCTS QUICKLY.

It takes the right collective mindset to make that happen. Finding the people who will make up your collective is the trick.

---

55 "The Purpose Economy," *Imperative.com*, https://www.imperative.com/purpose-economy/

THE SMART CREATIVE ARCHETYPE

"THEY ARE MULTIDIMENSIONAL, USUALLY COMBINING TECHNICAL DEPTH WITH BUSINESS SAVVY AND CREATIVE FLAIR. IN OTHER WORDS, THEY ARE NOT KNOWLEDGE WORKERS, AT LEAST NOT IN THE TRADITIONAL SENSE. THEY ARE A NEW KIND OF ANIMAL, A TYPE WE CALL A 'SMART CREATIVE,' AND THEY ARE THE KEY TO ACHIEVING SUCCESS IN THE INTERNET CENTURY."

ERIC SCHMIDT[56]

Google may have first coined the term *smart creative*, but it's become a ubiquitous catchphrase in the tech world. Creatives no longer refer to just designers. Visionary engineers, product managers, and even entrepreneurs can all be called smart creatives. These are the people you want to hire.

THEY SHARE SOME SPECIFIC CHARACTERISTICS SUCH AS:

- A user-focused point of view

- A flair for the hands-on — they don't just design things on paper; they jump right into prototype mode

- A willingness and enthusiasm to take risks, coupled with a company culture that encourages risk-taking, even when it leads to failure

- An autodidactical nature, ready to teach themselves a new skill if that's what it takes to get something done

- A disdain for the status quo, a blindness to organization structure, and a willingness to speak out in any role

THE DOWNSIDE OF SMART CREATIVES IS THAT THEY GET BORED EASILY.

To keep them engaged and happy, they have to be driven by and passionate about the purpose of your organization. When this happens, an employee will perform at a level far beyond someone simply working for a paycheck.

---

56 Eric Schmidt, *How Google Works* (Grand Central Publishing, 2014).

When the mission of a company is at stake, your team will self-manage and won't be as concerned with role definitions. People are far more likely to speak up and voice their opinions in this kind of environment — and that's good. We've seen some of the best engineering ideas come from non-engineers, and the environment needed for this is a common purpose and belief system where the team's clear objective is mission first.

THIS MINDSET IS PARTICULARLY CRUCIAL IF YOU'RE A STARTUP COMPETING AGAINST BIG TECH COMPANIES.

You probably won't be able to match the salaries they can offer with their deep pockets, and the best way to compensate for this is to double down on your purpose and what you aim to do with your product. That's why, when you scout and interview candidates, you need to be very open about your purpose, and ask the right questions to ensure that anyone you bring on board has a shared belief system. The best painkiller for the challenges associated with developing new products is to have a team with a unified purpose.

5.5 FIND YOUR SWEET SPOT WITH EMPLOYEE / CONTRACTOR BALANCE

Contractors provide flexibility. Employees are dedicated and reliable. Contractors do the job you hired them to do, do it well, and then get out. In a small company with an agile team, each employee will take personal ownership of the product and give you her blood, sweat, and tears. So how to decide which to hire, and how many of each?

The equation for determining how many employees versus contractors you need relies on three inputs:

1. COST
2. FLEXIBILITY
3. DEDICATION

LET'S GET INTO IT.

In our personal experience, the hourly cost of a contractor can be 30 to 40 percent higher than the hourly base cost of an employee. But looking at the base cost of an employee isn't an accurate comparable.

EMPLOYEES BRING ADDITIONAL COSTS, LIKE:

- Training and onboarding
- Medical, dental, and life insurance
- Paid vacation and sick leave
- 401(k) matching
- Workers' compensation
- And other benefits your company might choose to offer to entice full-time employees

When you do the math on all these other costs, an employee isn't always cheaper than a contractor. In fact, fringe benefit costs can be as much as 20 to 50 percent of the salary.[57]

You also have to look closely at the way employees and contractors work to gauge their cost. A full-time employee typically works forty hours a week. But 15 to 20 percent of that time is taken up with administrative tasks like timecards, corporate trainings, and other things that really don't contribute to your product's development. Depending on labor law in your state, you may have to pay overtime when an employee works beyond 40 hours a week. So to get the job done, you may actually be paying more than the base salary you intended on. And when things slow down, your employees are still expected to show up for work — and you're expected to pay for their time.

Contractors don't spend time on administrative tasks, and they tend to be more focused with their time in general, because they're working on specific areas of expertise and interest that they were recruited for. They have the potential to be way more efficient with their time than employees.

ONCE WE HAVE RULED OUT COST AS A FACTOR, WE ARE LEFT WITH A DECISION BETWEEN FLEXIBILITY AND DEDICATION.

---

57 Patrick Gillooly, "Should you hire a contractor or a full-time employee?," *TechRepublic*, September 19, 2000, http://www.techrepublic.com/article/should-you-hire-a-contractor-or-a-full-time-employee/

FLEXIBILITY

In the early stages of a company, contract employees can provide valuable flexibility. Contractors are fairly easy to hire, and don't require a lengthy, expensive recruitment process. If they don't turn out to be the right fit, they're also very easy to replace. And if the product pivots or even dies, having employees means firing employees. This means paperwork and liability. With contractors, this is not an issue.

DEDICATION

If the job requires a list of tasks to be completed quickly and well, then we're talking about a contractor. If, on the other hand, you're looking for someone to join the family, contribute to the organization's culture, and give you their all, then we are talking about an employee.

Weigh the pros and cons, then make informed decisions. And don't make them lightly. Hiring each employee should be a project. You're looking for someone who doesn't just have the right skills, but is going to potentially plus-one your culture. You're looking for someone who is dedicated to the success of the team, product, and organization.

A few pages ago, we talked about how small teams can be more productive during critical times of a business's growth. And along those lines, having a few talented developers can be wildly more productive than having a whole slew of mediocre ones. Your developers are your talent. They are not interchangeable. Being a good developer means being able to make intelligent, creative decisions that will affect how products work now and in the far-off future. Developers are coding artists with crucial skills.

> "IN SPORTS, THE EXISTENCE OF ELITE PARTICIPANTS IS WIDELY UNDERSTOOD AND ACCEPTED. IN PART, THIS IS BECAUSE SPORTS HAVE GOOD WAYS OF KEEPING SCORE. THEY HAVE COMMONLY ACCEPTED WAYS TO MEASURE THINGS. AT THE BARE MINIMUM, FOR EACH CONTEST, WE NEED A WAY TO KNOW WHO WON AND WHO LOST."
> ## ERIC SINK[58]

Imagine you were casting a movie. Your producer would never say "Just get me anyone with a SAG card; all actors are the same." The choice of actor can make or break a movie. Yet, when you're casting developers, you'll notice a lot of decision-makers saying "A developer is a developer. Just book one."

In our experience, a developer is not simply a developer. We were at WWDC (the Apple Worldwide Developer Conference) a few years ago talking to a technical director of software engineers. We asked him about productivity, and he impressed us by boasting that their top few developers working on new frameworks could accomplish as much as

---

58 Eric Sink, "Do Elite Software Developers Exist?," *Ericsink.com*, http://ericsink.com/entries/sports.html

twenty "normal" developers in the same amount of time. For one thing, without twenty different sets of opinions to contend with, they could get to work a lot quicker and get things done better.

KNOWING A GOOD DEVELOPER WHEN YOU SEE ONE
IS THE FIRST STEP TO BUILDING YOUR TEAM.

It's a task that should be going on all the time, and you should always be testing out new developers. Referrals are the best way to find these type of engineers. Another tactic we love is to hone in on projects that have been really successful and find out who was behind them. You don't always have to poach someone. Often, developers work on a freelance or contract basis.

Keeping good developers is the second step. Ironically, a lot of developers aren't simply motivated by money. Yes, you should pay them competitively; this goes without saying. But being inspired by your vision is what truly moves them. Do your research before you reach out to developers so you know what excites them — largely based on what type of projects they've historically worked on. But remember, they're not just going to want to make another of the same. They thrive on challenges, so your product needs to provide opportunities for them to improve their skills and learn new things.

While we're deconstructing the psyche of the developer mind, let us say something else: Developers can be quirky people. Understanding their quirks and listening to their needs is crucial to keeping them happy. And you want to keep them happy, because they're in high demand, and if they're not happy, they'll go elsewhere.

We think of developers as artisans with their own manifesto similar to ours. Git Tower published one that we think is the essence of how developers should think and act.

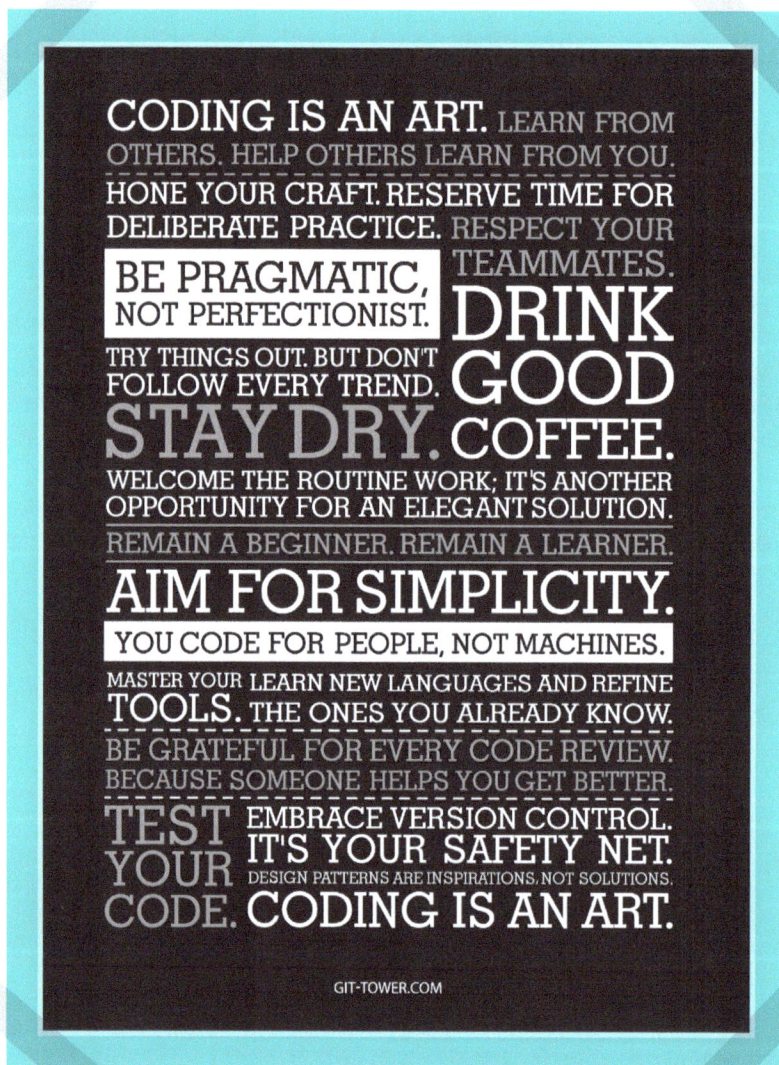

Image credit: https://www.git-tower.com/blog/developer-manifesto/

Just like any employee, developers thrive on frequent acknowledgement and praise, small rewards, and recognition of their value. But developers tend to be motivated and encouraged by weird and unexpected things, so taking the time to get to know them will get you far.

HERE ARE SOME REAL-WORLD EXAMPLES OF WAYS WE'VE SEEN DEVELOPERS BE MOTIVATED BY THEIR LEADERS:

- One leadership team bought their developers disc golf equipment and encouraged them to play an extended game once a week during lunchtime. The result? A happy team.

- Another encouraged a LAN gaming competition during lunch. Developers started inviting friends, and this fun way to blow off steam eventually became an effective recruiting technique. When the word got out about the lunchtime games, everyone wanted in.

- One of the developers on our own team loves trail running. A fancy new pair of shoes was part of his sign-on package. We've bought him a new pair every six months since. Yes, he could afford to by himself running shoes with what we pay him; that's not the point.

DIGITAL PRODUCTS DON'T LAUNCH WITHOUT THE ESSENTIAL TALENT THAT DEVELOPERS BRING. ONE OF YOUR TOP ROLES AS A COMPANY LEADER WILL BE TO FIND — AND KEEP — THE RIGHT DEVELOPERS.

5.7 MATCH YOUR DEVELOPERS TO YOUR TECH STACK

You've found a few stellar developers you're confident about. Now your challenge is to get them set up in the optimal environment for them to thrive. It's important to set expectations early on and revisit them often. At the same time, a talented developer needs a certain amount of autonomy. Striking the right balance is key.

ONE OF THE BIGGEST PRELIMINARY DECISIONS YOU NEED TO MAKE IS WHICH TECHNOLOGY STACK TO GO WITH.

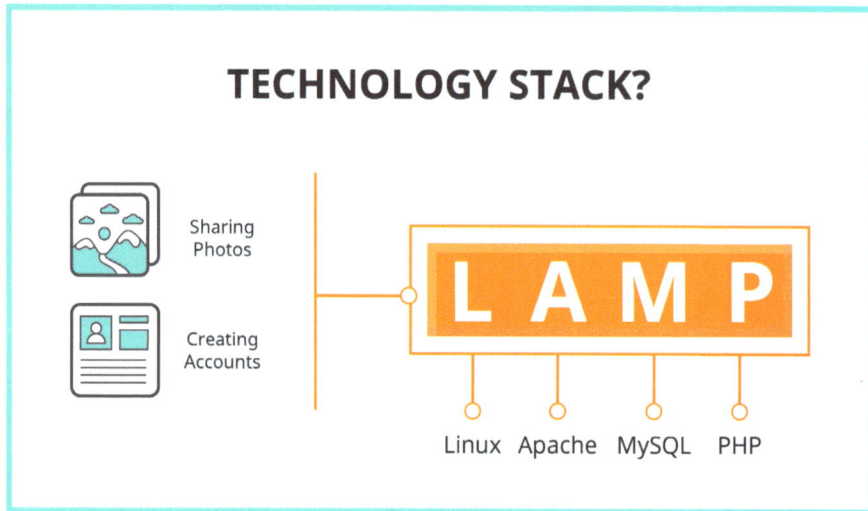

TECHNOLOGY STACK?

Sharing Photos

Creating Accounts

L A M P

Linux   Apache   MySQL   PHP

Too often, developers make product decisions based on the current buzz of internet forums and social media. Technology stacks are chosen based on biases and perceptions. And in nearly all cases, even when solid developers are on board, product decisions tend to be made not from experience but from what others are saying and from the example of miracle products on the market that don't represent the norm. If your developer has a tendency to harp on these unicorns or the opinions of so-called "thought leaders" in his or her arena, reign it in.

Right now is the perfect time to think about what technology foundation to start building on. You should never choose a development stack without building prototypes on key requirements and capabilities. Take a few days to test out some prototypes early on. Assuming your team has at least a few developers, have them split up and try out competing technologies. This can be the catalyst for good discussion and will be the seed of determining the best solution to go with. Always keep in mind though that user experience should come first, and the appropriate technology should follow. You are creating a user experience, not some awesome technology stack that you can brag about on forums after your startup runs out of money. Users don't care about your technology stack; they simply want your product to work.

Most development technologies arose to solve a niche problem. This concept sometimes gets lost on newer developers and even seasoned pros, but it's important to keep in mind so that you know which use cases to compare yours to when evaluating a technology stack. Every type of technology provides specific types of solutions and has its own attributes. In most cases, sticking with tested technologies that have been around for a while and are frequently updated will provide a number of benefits that newer technologies lack. For instance, a more established type of technology might have a large community with ample online help. When a problem arises, your developer can turn to this community to find solutions. Established technologies also tend to be more stable, because mature platforms use a release process that ensures good uptime records. And finally, established, frequently updated technologies are more secure. This is not always true, but often is.

If you did not hire developers from scratch, but instead are relying on a team within an existing corporate structure, you might be stuck working with what/who you have. In this case, it's important to be aware of whether these developers have a track record of delivering on time and meeting functionality goals. It's common for teams that don't deliver to blame their technology and request a new solution. Beware this phenomenon. The pattern we've seen again and again is that development teams use their technology as a crutch, but the real problem is a human problem that can't be solved by investing in different software. Excellent developers will deliver regardless of technology, so if you're finding a lot of blame placed on technology options, consider holding your developers to task to deliver in their current environment, and if that doesn't work, it might be time to look for some new developers.

HYPE-DRIVEN DEVELOPMENT

In 2016, Marek Kirejczyk of Daftcode wrote about the insidious phenomenon known as hype-driven development, where "Software development teams often make decisions about software architecture or technological stacks based on inaccurate opinions, social media, and in general on what is considered to be 'hot,' rather than solid research and any serious consideration of expected impact on their projects."[59] You can see this phenomenon in action after your developers get back from the latest tech conference with some bright, shiny new technology to rave about.

Hype-driven development is alluring, but it's not that effective, and can even be deleterious to your cause. As Kirejczyk says, "The problem with hype is that it easily leads to bad decisions. Both bad architectural decisions and technological stack decisions often haunt a team months or even years later. In the worst case, they may lead to another very problematic situation in software engineering: The Big Rewrite. Which almost never works out."[60]

The antidote to hype-driven development, according to Kirejczyk, is solid software engineering. Rather than focusing on what's trending on Twitter and Reddit, it asks developers to adopt solid practices when choosing technology: test and research technology options before you make a decision (which we, too, recommend). And Kirejczyk's advice is to vet developers so you're hiring people with strong technical backgrounds who understand various technology paradigms and have the perspective to make wisdom-driven choices rather than buying into hype.

To read more about the "Anatomy of Hype" as well as some examples of how hype-driven development can derail a product, Google Daftcode's blog post "Hype-Driven Development."

---

59 Marek Kirejczyk, "Hype Driven Development," *Daftcode*, November 23, 2016,
    https://blog.daftcode.pl/hype-driven-development-3469fc2e9b22#.n6lp188x4
60 Ibid

6.0 IDEAS THAT SELL ON EVERY FRONT

In the back of your mind, you're probably wondering

"HOW AM I GOING TO SELL THIS THING TO INVESTORS?"

Unless you're independently wealthy, you'll eventually need investors to take it to the next level. Obtaining investment funds is one of the first objectives of most entrepreneurs. But a lot of people fail because they don't know how to sell their idea. So how to build toward that eventual goal? In this chapter, we'll cover how to get there with a series of tasks that lead to the creation of a product deck.

Like every process we've advocated throughout this book, always create lean — at least in the beginning. During the initial selling process, you really just need a proof of concept on a napkin. Okay, the investor you're talking to might require more than a dirty scribble you just pulled out of your pocket. But seriously, don't over-deliver.

A BASIC, ONLY PARTIALLY FUNCTIONAL PROTOTYPE IS PLENTY.

Instead of building a fully functional product to bring in front of a VC, spend your time developing a solid product vision and doing some initial research (which you've already done, if you're reading this book sequentially). Use simple, basic tools to speed up your creative process. This isn't the time to impress anyone with your technology prowess. You can create a perfectly workable prototype in PowerPoint or Keynote.

ALWAYS USE THE SIMPLEST, CHEAPEST, AND EASIEST TOOLS TO CONVEY YOUR PRODUCT VISION QUICKLY.

If you are the sort who is thinking "I don't need anyone else, and I don't want to listen to what anyone else says; I'm just going to build this product myself," here's the truth: you're building a passion product. Certainly, there's nothing wrong with passion products, but that's not what this book is about. The end user of a passion product, for all practical purposes, is you, and you alone.

THERE ARE EXCEPTIONS TO THIS RULE.

In some rare cases, a passion product fully funded by the founder makes it to the mass market and is even successful. But it's the exception, not the norm.

Last note before we dive into your presentation prep. Although there are many cases of ideas invested in because of brilliant revenue models, ROI is not always the main motive for investors.

OFTEN, THEY ARE FAR MORE INTERESTED IN YOUR ABILITY TO SHOW ORGANIC GROWTH (THINK DAILY ACTIVE USERS) AND YOUR CAPACITY TO BRING YOUR PRODUCTS TO LIFE AND SCALE THEM TO A MASS AUDIENCE.

This book is largely about teaching you how to collaborate in order to create an innovative product that will change the lives of other people. Don't miss an opportunity to prove that mission — and your ability to act on it — to your investors.

A VALUE PROPOSITION IS A PROMISE OF VALUE TO BE DELIVERED, COMMUNICATED, AND ACKNOWLEDGED.

At the same time, it's a belief of the customers about how that value will be delivered, experienced, and acquired.

Consumers spend a lot of time looking around for the best value from products. Your value proposition clearly identifies the benefits you provide, to whom, and how. It's a creative statement that depicts your unique selling point. Your value proposition should clearly identify the benefits consumers will get from your product, why it's better than a competing product in the same market (a.k.a. perceived substitutes), and how the price compares.

AN AWESOME VALUE PROPOSITION IS A REASON TO INVEST IN YOU.

At its most basic level, your value proposition should answer these questions. The answers will help your product stand apart from the alternatives vying for angel dollars.

WHAT IS YOUR PRODUCT?

HOW WILL YOU DIFFERENTIATE IT?

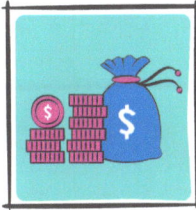

HOW MUCH WILL IT COST?

HOW WILL IT SOLVE A PROBLEM FOR USERS?

WHERE WILL YOU SELL IT?

AND WHO ARE THE USERS IT WILL SOLVE THE PROBLEM FOR?

HOW WILL YOU PROMOTE IT?

But, here's the catch: your value proposition must go beyond simple answers. It must also include a compelling statement that points out your unique benefits — things that make you far superior to even your closest competition. By defining and communicating your value effectively, you don't just sell well; you create a personal and honest relationship with your target audience — the real reason they will choose you.

That's why we advocate for a process of establishing a value proposition that takes into consideration not just your product (the "what"), but your brand (the "why"), and the user experience you're creating (the "how") — and not necessarily in that order.

"MANY ENTREPRENEURS LOSE OUT, DUE TO NEVER TRULY ARTICULATING A COMPELLING VALUE PROPOSITION."

MICHAEL SKOK, FOUNDER STARTUP SECRETS AT THE HARVARD INNOVATION LAB[61]

To help you zero in on your product's value proposition, it's helpful to do an exercise on paper we call the BxP Value Proposition Map. This is an excellent way to understand the underlying motivations of your potential users and why they tend to choose a particular product to solve a problem, save them time, or eliminate an inconvenience. It's also a good way to get honest with yourself about the competition, and compare your value to theirs. Remember, unless you have something novel and unique to offer, you won't be competitive in the eyes of investors.

To understand just how well your product is differentiated, there are three things you want to consider:

1. Brand (yours)
2. Experience (of your users)
3. Product

Get out a piece of paper and a pen. Make a sketch that looks something like this:

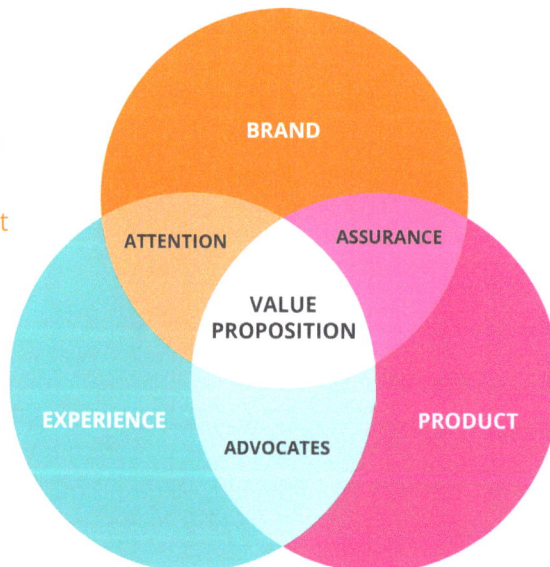

BRAND

ATTENTION   ASSURANCE

VALUE PROPOSITION

EXPERIENCE   ADVOCATES   PRODUCT

(Or use the template in our product workbook, *Got Ideas? Work Them Out Here*)

---

61 Michael Skok, "4 Steps to Building a Compelling Value Proposition," *Forbes*, June 14, 2013

NOW, BRAINSTORM.

Users develop emotional connections to brands that far supercede any alliance to features, yet "brand" is not something most entrepreneurs focus their efforts on when they're head-down creating the actual product. That's a mistake. Brand is important to think about from the very beginning, and it's certainly crucial to your search for funds.

The reason we consider it the first part of the BxP equation is quite simple: When it comes down to it, brand is the number one indicator of value for a customer. Apple may have a legendary product with the iPhone, but we'd argue that the actual product is less persuasive to buyers than the Apple brand.

Brands are often built over time. A lot of the digital brands you recognize today took years to get to this point and burned up lots of seed money before they entered mainstream awareness. Even Apple was once a niche brand.

If you're developing a product no one has ever heard of before, you're automatically at a disadvantage when competing against well-known brands. People tend to choose brands they know to solve their problems. If they trust Tide laundry detergent, and Tide comes out with an eco-friendly detergent, they'll buy that over a startup brand they've never heard of. This actually happened by the way. In 2016, Tide announced their new eco-friendly detergent, Tide purclean™.

PEOPLE ALWAYS OVER-VALUE A PRODUCT THEY CURRENTLY LIKE. YOU HAVE TO WORK EXTRA HARD, AS A NEW BRAND, TO CONVINCE PEOPLE OF YOUR VALUE.

This does not mean you have to drop everything to go throw tons of money at branding. For now, it just means you have to work your vision for your brand into your value statement. (Our second book, *Got Users? How to Persuade People to Not Just Buy But Love Your Product*, goes deeply into detail about branding.)

Your brand is something you'll build concurrently with your product, so by the time you get to creating a pitch deck, you'll perhaps already have a logo and some design elements in place. But brand goes far beyond visual identity.

ON A DEEPER LEVEL, IT'S ABOUT HOW YOUR PRODUCT MAKES PEOPLE FEEL, AND WHAT ASSOCIATIONS THEY WILL HAVE WITH YOUR COMPANY AND PRODUCT.

It extends across every interaction they have with you at every touchpoint.

So as you're thinking about your value proposition, ask yourself two very important questions right up front:

HOW WILL YOUR BRAND MAKE PEOPLE FEEL?
AND WHAT ASSOCIATIONS WILL THEY HAVE WITH IT?

In or next to the "brand" circle on your map, write down a few words about this.

EXPERIENCE — HOW PEOPLE EXPERIENCE YOUR PRODUCT

The experience of your users is key to reinforcing brand perception because it connects your product's utility with your users' emotional reaction. Experience encompasses everything from onboarding to social media presence to customer service.

User experience is the catalyst for creating brand advocates, who then create brand awareness for your new product. Ultimately, it's not enough to just understand "why" someone would want your product; you have to understand how they will use it in order to understand its true value to them.

In or next to the "experience" circle on your map, write down the user-experience angles that will differentiate you from competitors. What will set your user experience apart and "delight" your users, in the current parlance of the times (although it's really timeless)? Forget about the generic stuff like "We'll have excellent customer service." Most people won't believe you when you say that, anyway. Write down just a few words specific to things you can and will do better and differently.

PRODUCT — WHAT UNIQUE WAY YOU'VE SOLVED A REAL USER PROBLEM

Now for your actual product — ironically the very last piece to your value proposition. What are its key features and benefits? What user pain points does it solve?

Your product might simply do something ten times better than the products before it did. It might solve a problem that hasn't been solved yet. Or it might address an untapped need in the market. Do your homework so you're confident about the fact that you are, indeed, novel.

In or next to the "product" circle on your map, elucidate what makes your product itself truly different. What core features align with user needs and solve user problems? What features differentiate it from other products in the market that it may compete with?

Your goal should be to create a product that does something so natural/special/unique that your users can't help but talk to their friends about it. Nearly all of the current lineup of successful startups grew based on word of mouth, so marketing and sales budgets were largely redirected to product and experience development.

Once you've taken notes on the areas of brand, experience, and product, spend a little time whipping your ideas into a cohesive value proposition that encompasses all three.

The purpose of this exercise is to figure out where you are superior. You can't be just as good as your competitors, or even a little bit better. You have to be vastly superior in one or more of these categories.

WE CAN'T STRESS THIS ENOUGH: THE SUCCESS OF YOUR PRODUCT IS DIRECTLY TIED TO THE VALUE YOUR CUSTOMERS GET FROM IT.

The value that customers believe they get from your product drives every decision you make: marketing activities, production choices, and investment decisions.

Keep in mind that customer perception — along with customer desires — can change. And value propositions can certainly vary across industries and different market segments. This can put a lot of pressure on companies to constantly invest resources in market research in order to keep up with fickle customer moods and dynamic markets. But creating and delivering on value proposition is still a universally important step for business founders.

WHERE BXP VALUES COLLIDE

When you create a value proposition well-rounded by effort to carve out your Brand, Experience, and Product, the payback lies "between the lines." The assets your efforts beget are Attention, Assurance, and Advocates.

It's these assets that create true value for your product, and why it's so important to pay attention to all three of the main values as you build out. Even a startup in scramble mode should actively work toward creating value for users in all of these areas from the beginning.

BRAND + EXPERIENCE = ATTENTION

AT THE CONFLUENCE OF BRAND AND EXPERIENCE IS ATTENTION.

The digital product landscape is both massive and fragmented. There are hundreds of millions of apps and services, so earning the attention of users is immensely challenging. It takes a strong, compelling, relevant, authentic, and relatable image.

As we mentioned already, existing brands with a high level of credibility among consumers already have the advantage of being viewed as trustworthy. You don't have that advantage, so you must build it with a carefully calibrated effort to carve out a brand personality in tandem with an incredible customer experience. If you do a spectacular job of this, you can gain attention from potential users without massive ad spend or endorsements.

BRAND + PRODUCT = ASSURANCE

AT THE INTERSECTION OF BRAND AND PRODUCT IS ASSURANCE.

Once you have their attention, you must assure them that you're the right product for their needs. This quality is important both for attaining new customers and keeping the ones you already have.

If 80 percent of your product's usage will likely come from 20 percent (or less) of your users, it's imperative to continually reassure your existing customers that they made the right choice by going with you. This reassurance comes from your excellent product and, perhaps more importantly, from your seemingly benevolent brand.

And, of course, as you're building your customer base, you will also need new customers — lots of new customers. Assurance is a differentiator when there are many compelling options with similar features, specs, and functionality. Your product must be the safer, better bet.

EXPERIENCE + PRODUCT = ADVOCATES

Without unlimited marketing dollars, the only true way to organically grow a product is to have a set of advocates that tell their friends, family, and contacts about your product.

WHERE EXPERIENCE AND PRODUCT INTERSECT, A RESIDUAL "SIDE EFFECT" IS PRODUCT ADVOCATES.

This includes your power users, but is not limited to them.

The value of one-to-one testimonials of product advocates is hard to measure, but it can truly be the gasoline that fuels rapid and continued growth. And the best way to cultivate product advocates is with a product that gives an excellent experience.

Here's the catch about this BxP exercise: it's just an exercise. The real work obviously comes from your product, brand, and company build-out. But this exercise can help you elucidate where you're trying to get to and why it's so important that you pay deep attention to all three values:

BRAND, EXPERIENCE, AND PRODUCT (IN THAT ORDER).

6.2 BRING YOUR BUSINESS MODEL TO LIFE

WE ALL ENJOY STORIES. THAT'S PART OF BEING SOCIAL CREATURES.

A business model tells a story in a very specific way. It's a design for how you intend your product to innovate, your people to thrive, and your business to scale. With potential investors as a rapt audience, you'll spin the true tale of how your product will come to life, change history, and make everyone rich — or some toned-down version of that product utopia.

At this point in your product development, you probably don't have the time or information you need to create a big fancy, Harvard-approved business model. But you do need to start thinking about it so that when you're working on your pitch deck and standing in front of investors, you sound like you do have a plan.

In the last section, we introduced the idea of a value proposition map, and as you think about your business model, the ideas you generate should build on those discoveries.

And like the value proposition map, it should hold the user at the center of every decision. It should also illustrate how your product will generate value for customers and revenue for owners and investors. Your business model takes your value proposition and adds on a tactical plan. That plan should summarize, with a certain amount of detail, how you'll run and grow your business. It should prove that your current company objectives and goals map to growth.

In this section, we'll run through an exercise that covers three main topics of business models: users, assets, and a profit formula.

You may be creating a business model with a particular investment meeting in mind. But, as with every exercise in this book, you'll revisit it as your product and company evolves.

IN FACT, YOUR BUSINESS MODEL MIGHT NEED TWEAKS EACH TIME YOU GO THROUGH AN ITERATION OF YOUR PRODUCT USING THE MAKE/SHOW/MEASURE/LEARN/REACT CYCLE.

UAP: THE ELEMENTS OF A MODERN BUSINESS MODEL

When it comes to creating a business model, we believe in a formula we call UAP: Users, Assets, Profit Formula. It encompasses the things investors want you to prove you have a handle on.

As your company develops, your leadership will come to represent these three areas as well. Eventually, you'll have someone in charge of user experience, someone in charge of your assets, and someone in charge of your finances and revenue model.

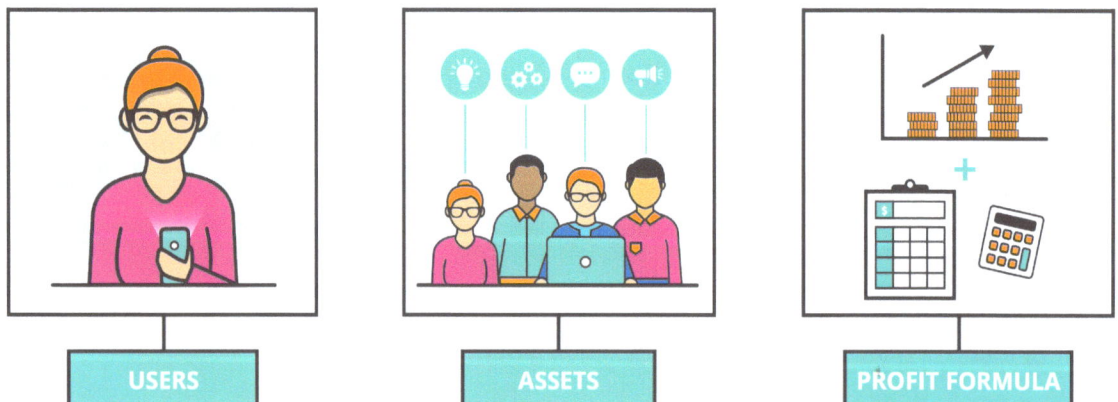

USERS

ASSETS

PROFIT FORMULA

USERS

At the center of creating a business model is your steadfast dedication to a truly customer-centric business. Your business model should include tangible information about:

- Who your target users are (your customer segments)
- What you have to offer them (your value proposition)
- What channels you'll find them on
- How you'll get them to convert to customers
- Evidence you have found at least a few users who love your product

It's not just about creating customers, of course, but retaining customers with a relationship strategy throughout their customer lifecycles, and even converting them into brand advocates. Your business model should include descriptions of the channels, platforms, and partnerships you'll employ to move users through the process of becoming advocates.

Your business model should include the key performance indicators (KPIs) you'll use to measure success in acquiring, retaining, and converting users. These KPIs will map out your path to user and revenue growth, which is exactly what investors care about.

ASSETS

Your business model should also outline the key people and technologies required to fulfill your value proposition and meet your financial and user goals. Your people and the development model you choose will both play a key role in determining the viability of a product and the ability of your product to continuously iterate and improve.

- What key activities do you need to do in order to be able to perform?
- What key resources are indispensable to your business model?
- What partnerships can you leverage? — for instance, cloud content platforms or third-party APIs you can integrate so you don't need to build everything from scratch.

Put some thought into your tech stack, the custom or proprietary technology solutions you'll require, whether this stack will scale, and how much it will cost.

PROFIT = TOTAL REVENUE - TOTAL COST

Obviously, no matter who you're showing it to, your plan for making money is crucial to a business model. When charting your profit formula (a.k.a. revenue model), consider both your revenue streams and your cost structure.

But you don't have to get bogged down in spreadsheets or enlist a numbers genius to put this on paper. Complicated formulas and scenarios are unwieldy, expensive, and require lots of resources. A few sentences might be adequate.

CASE STUDY: AIRBNB

Talk about disruptors. When Airbnb was dreamed up by a trio of founders in 2007, not many would have predicted that the wacky idea of letting people pay to stay in strangers' houses would soon upend the hotel industry.

But a few early visionaries had enough faith to throw money in the ring — a lot of money — which helped grow Airbnb into the household name (pun intended) it is today.

Here's how Airbnb's business model helped sell an idea to the people who could make it happen.

- **Users**: The startup envisioned travelers connecting with local hosts for economical places to stay. The traveler gets an adventure; the host makes money. Over the years, this model has grown to include business travel and other types of short-term stays.

- **Assets**: The key to making the user model work was technology. Airbnb is driven by a user-friendly website and app that allow both travelers and hosts to "self-serv." This positioned Airbnb to build a highly scalable model that to date has seen over 300M+ guest bookings.[62]

---

62 "Fast Facts," Press.airbnb.com, accessed August 31, 2018, https://press.airbnb.com/fast-facts/

- **Profit Formula**: Airbnb proposed a very simple business model in their initial pitch:

WE TAKE A 10% COMMISSION ON EACH RENTAL.

It doesn't have to be fancy. In fact, simple is often better.

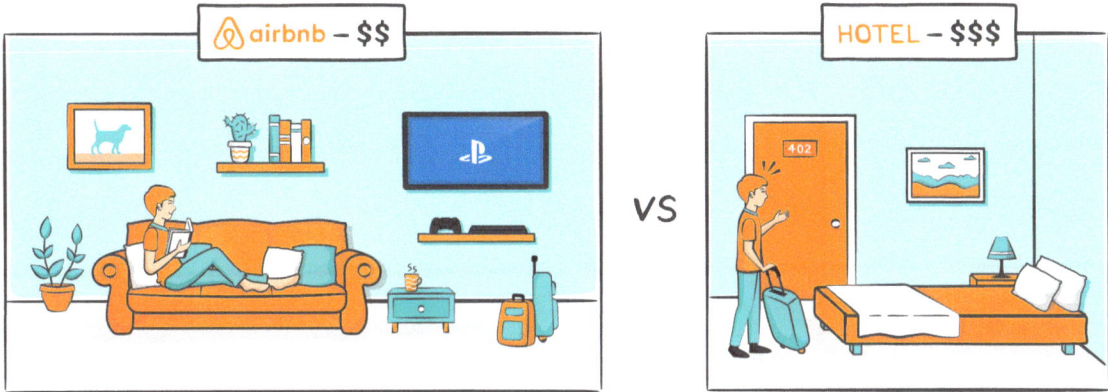

HOW TO PROVE YOUR BUSINESS MODEL

The best way to prove your business model over time is to set it up with a number of KPIs that will prove the status and health of your company as time goes on.

HERE'S OUR TOP TEN METRICS:

1. ACTIVE USERS (DAILY AND MONTHLY)

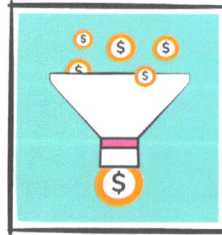

4. CONVERSION RATE

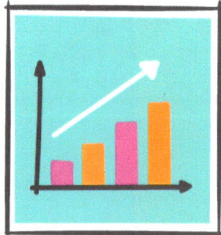

2. MONTHLY GROWTH (NEW AND ACTIVE USERS)

5. CHURN

3. BOUNCE RATE

6. CUSTOMER ACQUISITION COST— PAID AND BLENDED (CAC)

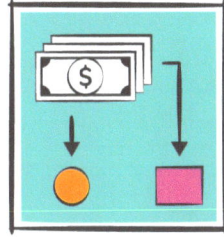

9. GROSS AND NET BURN RATE (CASH FLOW)

7. GROSS MERCHANDISE VALUE (GMV)

10. CUSTOMER LIFETIME VALUE (CLTV)

8. COMMITTED MONTHLY RECURRING REVENUE (CMRR)

Your key KPIs should have a direct connection to your profit formula so that the status and health of your product can be clearly measured as time goes on. It's important to have KPIs that allow you to assess the health of your company early on and at any stage, so you can assess and reassess whether your profit formula actually works.

IF YOU'VE DONE SOME SOUL SEARCHING, IT'S TIME TO GET IN THE RIGHT FRAME OF MIND TO PITCH TO VCS.
THAT IS, THE FRAME OF MIND TO DO SOME FRAMING.

6.3 KNOW YOUR AUDIENCE: THE FINE ART OF FRAMING

"FRAMING BIAS SUGGESTS THAT HOW SOMETHING IS PRESENTED (THE 'FRAME') INFLUENCES THE CHOICES PEOPLE MAKE."
FRAMING AND FRAMING THEORY[63]

THE TERM FRAMING COMES FROM COGNITIVE SCIENCE, WHICH EXPLAINS THAT PEOPLE MAKE DECISIONS BASED ON HOW CHOICES ARE PRESENTED TO THEM.

When someone says "tiger," we automatically think of things like big cat, teeth, stripes, etc. Each of these associations in turn have secondary associations. The idea of a tiger is both thrilling and terrifying.

When someone tells you they have "a great new idea," what associations come to your mind? The range of possibilities goes from investment opportunity to potential wealth to inevitable failure. Similarly, when you propose a new product idea, certain associations come to mind for your audience.

---

63 "Framing and Framing Theory," *California State University Northridge*, accessed July 7, 2017, http://www.csun.edu/~rk33883/Framing%20Theory%20Lecture%20Ubertopic.htm

Getting to know your audience will help you predict their frame. If your audience believes that your new idea is an investment in their future or their company, they may be more inclined to believe the numbers in your presentation. Conversely, if a "new idea" conjures associations like risk, failure, cost, or liability, they will question your facts and figures and be less likely to believe them.

You can use framing to your benefit. When the frames are in alignment, your audience will take in your facts and figures and be more likely to trust they are accurate — upping your chances of getting the approval to move forward and work out a deal.

If at any point during the presentation it becomes clear that a person's frame is out of sync with your own, it's time to abandon facts and figures immediately and reframe the discussion or presentation. This can be a very challenging process, and sometimes it's best to simply abort mission and find a new audience. The effort you'll have to put into trying to reframe for this one probably won't be the best use of your time.

WHEN IT COMES TO FRAMING YOUR PROJECT FOR INVESTORS, ONE OF THE MOST ELEMENTAL CONCEPTS TO KEEP IN MIND IS STATUS.

STATUS AS A SELLING POINT

Status is how a person, product, or company measures up to others doing similar things. Knowing how to wield and work with status — your own, your product's, and the investor's — can prompt easy agreement or trigger envy and frustration that can lead to failure to sell your idea. Learning to negotiate around status is crucial to embolden your audience and sell your idea.

Before you get in front of an investor audience, do your research. Here are the important things to know:

KNOW WHAT KIND OF INVESTMENTS THEY LIKE.

If you're pitching to an investor, it's a given that you're trying to show how financial reward will offset financial risk. Go beyond this, and differentiate your idea by describing how the investment will pay off in additional dimensions beyond just the financial.

Status is very important to investors. Offering them an increase in their perceived clout is a smart way to get funded. In fact, it can be even more important to investors than financial returns.

Everyone in Silicon Valley knows the name Sequoia Capital because that firm has funded some of the biggest hits in history: Apple, Google, Oracle, PayPal, YouTube, Instagram, Nvidia, LinkedIn, and Yahoo. Sequoia Capital doesn't enjoy the ego boost of all these successes; such wins help them stay relevant, so the best ideas come to them first. One of the things you might have to offer an investor is an inroad to a new market. You could be the perfect product to help them expand their portfolio and increase their credibility.

By getting to know your audience better, you will gain an idea of how to increase their perceived status and help them build more credibility with their peers. If your presentation can show them a clear path to improving their status, and explain how they'll get a win from their partnership with you, you establish credibility — and that creates an enormous desire.

LEARN FROM OTHERS.

Before you present, it's a good idea to seek out others who have already presented to this same audience — specifically, others who have gotten funded by the investor in question. They've already navigated these waters, and obviously made a worthy impression. How did they do it?

People who have succeeded at their dreams love talking about their methods. They're eager to share their experiences and stories. Reference these stories in your own presentation, and you automatically elevate your status. By understanding what others went through, you're setting yourself up to gain product approval and funding under similar circumstances. And when you talk about the powerful lessons you've learned from others during your presentation, it shows passion, commitment, and a level of earnestness.

You can usually see a list of boasts on a VC's website, but another trustworthy resource for this sort of information is Crunchbase, which lists which VCs have invested in which startups.

BE REALISTIC.

You might have a clear notion of why your product is better than others, but that connection may be harder than you expect for your audience to understand. As you're building your presentation, constantly keep your specific audience in mind. If you're comparing your brand-new product idea to an established brand, don't just assume that it's a useful comparison. You don't yet have any active users; they're a $1B unicorn. Is your comparison realistic? Or are you reaching for the stars?

It's interesting to note that research shows men and women view comparisons differently. Male and female brains have actual structural differences — the female brain tends to have stronger connections between hemispheres, putting language and memory in touch with emotions. Female brains also have a larger hippocampus, which makes for better recall of detail. Male brains, on the other hand, tend to enable superior spatial ability and problem-solving processes. In marketing, this means that women are more likely to respond to emotional cues and detail; for men, keep it simple but well-designed."[64]

BUILD TRUST.

Trust is earned over time. If you're seeking funding from someone you already work with regularly, you have an advantage because you've had a chance to develop a trustworthy relationship. Let's phrase this a different way: if you think you might be asking someone you already know for funding, start building a foundation of trust way in advance. Then, when it comes time to make a big ask for funds or seek approval on a project, you can leverage this trust to make the deal. This sort of trust can be built over time with baby steps and through any other work you've done outside of this project. Your audience will feel better and give you more leeway when there is an established relationship of trust.

Of course, you're typically approaching investors you've never worked with before or even met, so you won't have a chance to establish a trust rapport in advance. If you've successfully brought other products to market, or been part of a team that has, super. Include those boasts. It implies trustworthiness and capability.

But if this is your very first product ever, your best bet for getting a potential investor to trust you is to have a solid core of engaged users for your fledgling product. This "proof in the pudding" pretty much speaks for itself.

Knowing who you're pitching to is crucial, but exactly how do you tell your product story to investors and everyone else at this early stage of the game, long before you have presumably hired any marketing experts?

Most people do it with a product deck.

---

64 Jenn Fusion, "Marketing to Men Vs. Women," *Chron*, accessed July 7, 2017, http://smallbusiness.chron.com/marketing-men-vs-women-1011.html

HOW MUCH SHOULD YOU SHARE

There's a paranoia in the inventor world that if you share your idea before you've had a chance to create, copyright, and patent it, someone might steal it. We've all seen *The Social Network*. IP theft is real.

Still, ideas are not rare. Most visionary people have so many ideas that there's no way they could get to them all in a lifetime. Finding the time and resources to bring them to life is the hard part. Most of the people who have the resources and can-do spirit to birth a product already have their own ideas. They don't need yours.

Don't believe us? Just ask Elon Musk. He's had some of the best ideas in modern times, and he's built entire companies out of them. Take Tesla Motors, the electric-vehicle company he has built into a major player in the auto industry, with one of the hottest commodities on the market. Musk did a bold thing a few years ago: he took all the company's patents down off the wall of its Palo Alto HQ and publicly declared:

TESLA MOTORS WAS CREATED TO ACCELERATE THE ADVENT OF SUSTAINABLE TRANSPORT. IF WE CLEAR A PATH TO THE CREATION OF COMPELLING ELECTRIC VEHICLES, BUT THEN LAY INTELLECTUAL PROPERTY LANDMINES BEHIND US TO INHIBIT OTHERS, WE ARE ACTING IN A MANNER CONTRARY TO THAT GOAL. TESLA WILL NOT INITIATE PATENT LAWSUITS AGAINST ANYONE WHO, IN GOOD FAITH, WANTS TO USE OUR TECHNOLOGY.[65]

Musk also said that "technology leadership is not defined by patents," and we agree. A commitment to open-source code, after all, is the foundation of the internet itself.

So don't be afraid to share your ideas with those who can help you realize them.

---

65 Elon Musk, "All Our Patent Are Belong to You," Tesla.com, June 12, 2014, https://www.tesla.com/ blog/all-our-patent-are-belong-you

172

6.4 CREATE A PITCH-PERFECT PITCH

The fine art of storytelling helps you humanize your product and create an emotional connection. Ideas radiate from creative teams, colliding with other ideas to generate even more inspiration. As you filter through all of these inspired ideas, you start to craft a compelling story that your audience can follow along with you.

"SCRATCH THE SURFACE IN A TYPICAL BOARDROOM, AND WE'RE ALL JUST CAVEMEN WITH BRIEFCASES, HUNGRY FOR A WISE PERSON TO TELL US STORIES." — DR. ALAN KAY[66]

Your product deck is a visual presentation, a snapshot that helps you sell your idea at a high level — to investors, potential talent, and any number of other stakeholders.

Product decks will vary in style, length, and the actual elements they contain, but there are some things that tend to be consistent from deck to deck. Rather than simply define each of these things, let's look at a real-life example of a company that pitched to a major VC firm, won, and went on to become one of the true internet greats: YouTube.

---

66 "YouTube by the numbers," *YouTube*, accessed July 7, 2017, ttps://www.youtube.com/yt/press/statistics.html

Today, YouTube has over a billion users and reaches more millennials than any cable network in the U.S. But once, YouTube was just another startup in need of seed money. The fledgling company got this funding from Sequoia Capital, whose own tagline, by the way, is a wonderful pitch statement:

WE HELP THE DARING BUILD LEGENDARY COMPANIES.

Indeed. Anyway, thanks to Miles Grimshaw, currently an investor at Thrive Capital, you can see YouTube's original deck, along with the memo that Sequoia partner Roelof Botha circulated internally encouraging the VC's decision-makers to invest in YouTube in these early days. What we love about YouTube's deck is its extreme simplicity. Here's how they did it in their original pitch deck, way back in 2005.

LOGO AND TAGLINE

Your branding may change as you get funding that you can throw at an expensive firm one day, but in the beginning, even a rudimentary effort at a logo and tagline can go a long way in making people remember you. A strong brand identity helps you stand out in a densely crowded world of endless products and offerings, and will fuel recognition and recall with everyone you interact with.

You Tube
Broadcast Yourself ™

YOUR PURPOSE

A single declarative sentence describing why you are creating this product. This is like an elevator pitch: It's one sentence long. It fits on one slide. It's simple and captivating, grabbing your audience's attention immediately.

Here was YouTube's:

Company Purpose

To become the primary outlet of user-generated video content on the Internet, and to allow anyone to upload, share, and browse this content.

THE PROBLEM

Begin your story with a problem that needs to be solved. Describe the pain of your customer, and how this customer is currently dealing with the problem. Your goal is for your audience to understand the way you think in a way that sparks empathy within them.

Problem

- Video files are too large to e-mail.
- Video files are too large to host.
- No standardization of video file formats.
- Videos exist as isolated files.

THE SOLUTION

Now, take your audience on the journey of how your product will solve this problem. Explain in simple, clear terms how your product makes the customer's life better.

Solution

- Consumers upload their videos to YouTube. YouTube takes care of serving the content to millions of viewers.
- YouTube's video encoding back-end converts uploaded videos to Flash Video.
- YouTube provides a community that connects users to videos, users to users, and videos to videos.

THE MARKET

Define your market, including its size. You should be able to verify your numbers with data.

YouTube didn't do this, but you could also outline a timeline of events leading to the evolution of your product category. This should conclude with why your product is relevant right now and why there is urgency around building it.

Market Size

- Digital video recording technology is for the first time cheap enough to mass-produce and integrate into existing consumer products.
- Broadband Internet in the home has finally reached critical mass, making the Internet a viable alternative delivery mechanism for videos.

THE PRODUCT

Of course, it's often a good idea to give your audience a glimpse at how your product works. Be careful though, because the saying "you live and die by the demo" is true. Conducting a live product demo in front of investors is just asking for trouble. Instead, we recommend you screen-record or otherwise "fake" the demo. Otherwise, you might end up beset with presentation hurdles like resolution issues on a foreign projector, computer adaptor snafus, network connectivity trouble, etc. You can assume from looking at the YouTube slide below that the person who presented this deck probably backed it up with more in-depth verbal explanation, but these are some good bullet points.

Product Developmet

- Community
- Open architecture
- Target vertical markets with a need for video content
- Features currently in development

THE BUSINESS MODEL

Show how your product will make money. You can co-opt some of the work you did in 6.2. Talk about where you are going with your product, how you expect to get there, and why you believe it can be successful.

Sales & Distribution

- Advertising
- Act as a for-pay distribution channel for promotional videos
- Charge members for premium features
- Charge viewers for premium content

KEY PERFORMANCE INDICATORS (KPIS)

KPIs are metrics you use to evaluate the success of your business according to goals you define early on and revisit as you go. They might include things like your net revenue or customer retention. By setting them up front, you can then look for the right data to prove how you're performing against them.

Metrics

- Launched June 11th. Has already overtaken all previously existing competitors and is now the dominant player in this space.

THE COMPETITION

Here, start by listing your competitors, then explain why you're better than them. Why is your product different? Be explicit. This is a place for detail. If you understand the existing ecosystem of products and have insights, this is the time to show your breadth of knowledge and acknowledge the challenges that lie ahead. Show that you have the determination to see the product or idea through, using other examples, past performance, insider knowledge, etc. Show that you are adaptable to a changing environment and quick to adjust on the fly. You will be constantly tested when developing a new idea. Illustrate that you are able to handle tight situations under pressure.

Competition

— OurMedia.org, Open Media Network, Google Video

— PutFile, DailyMotion, Vimeo

THE TEAM

Introduce your team. Explain why they are awesome, where they came from, and why they will stick together to deliver on the product. YouTube's founders specifically referenced their experience at PayPal to create status and show a track record of success and experience.

Team

— Steve Chen: Recruited by Max Levchin as one of PayPal's first engineers; University of Illinois, Computer Science

— Chad Hurley: PayPal's first designer, responsible for PayPal logo, main features, and design

— Jawed Karim: CS Graduate student at Stanford University; Recruited by Max Levchin as one of PayPal's first engineers; University of Illinois. Computer Science

(Thanks to Miles Grimshaw for posting YouTube's original deck along with the internally circulated memo about it on his website, milesgrimshaw.com.)

There are a few other elements we often recommend entrepreneurs and product makers put in their decks, depending on the stage of the project and the intended audience. These include:

THE MARKET STRATEGY

Show how you will acquire users and via which channels. Speak to your cost per user acquired. Define your key performance indicators. Show the performance of the KPIs. Outline what you project future performance to look like.

SUPPORTING DATA

Related to market strategy, some data points about the existing market — how many people use apps of a certain type, for instance — will show that you have conducted the proper research, collected a ton of insight, and used this knowledge to create a detailed and informed plan.

THE ASK

All of this groundwork leads you to "the ask." You've prepared your audience and paved the path. Now it's time to clearly outline what you are asking for. Keep it simple, but don't be afraid to ask for exactly what you need. Is it money? Resources? An excellent engineer to join your team? Spell it out, and be bold.

AND HERE'S A TIP: ASK FOR A LOT MORE THAN YOU THINK YOU NEED.

Digital products are notorious for hidden surprises that inevitably add to the time and cost to ship. No matter how well you plan, even for people with years of experience, life as an entrepreneur in the digital space will place surprises in your path every day. The faster you're moving, the faster they will come.

If you're building with the RAD model (see 3.3 Prototype Faster, Learn Better: RAD), you can't predict the state of your app a few months from now. Estimating the cost will be educated guesswork at best. In addition, when you're reacting quickly to user feedback to iterate product revs, you'll make little mistakes. You'll also uncover opportunities you can react to, and chances to pivot. This is all good, but it can require funds you didn't plan for. In the big scheme of things, these small early mistakes will pay for themselves. But right now is not the time to be conservative. Take whatever number you think you'll need and double or even triple it.

The purpose of getting VC funding is to provide rocket fuel for your launch. You'll want enough cash to be able to do things like:

- Work in multiple locations (maybe you'll fly across the country to be with your users in person, like Airbnb did in the early days)

- Explore a better product/market fit with a derivative product

- Forge an exclusive partnership with a much larger company that requires you to ramp up without delay

ANYTHING COULD HAPPEN. THE KEY IS TO REMEMBER THAT YOU WILL NEED MORE MONEY THAN YOU THINK YOU DO.

This might not be a slide in your deck (it wasn't in YouTube's). It might be a verbal request. But every presentation needs a "call to action," so to speak.

Your entire deck should illustrate the reward your product will invoke on whomever is involved in its production. In order to do so, you must deeply understand the needs and desires of your audience. They might be looking for clout, or money, or a promotion. Ensure you show a clear path to the result they want.

Storytelling is key to successful presentations. But stories don't have to be complex. In fact, a short, well-told story often has more positive impact than one laced with too much detail. Of course, it can be harder to create a beautiful 10-to-20 slide presentation than one with 50+ slides. Knowing what to leave in and what to take out takes practice. You can learn by watching others with experience present — and noticing how engaged their audience is.

Always remember that your audience is probably experiencing fear of the unknown. They are at the threshold, yes, but still asking themselves, "How much is this going to cost? How will this affect our team? How can we do this with all of the other competing priorities?"

Additional reassurance is needed quickly and succinctly. You want to make it personal for them so it becomes a partnership to solve the problem you've outlined. This can be accomplished through a variety of methods, but the top three are:

1. A recommendation from an industry-recognized thought leader validating your idea

2. A comparison of your product and similar successful product case studies, making sure your product is unique and your comparisons illustrative

3. Thoughtful and relevant user research

You can also combine some of the methods above for a more holistic approach to reassurance.

Prepare your audience for what's to come. You want to get your idea built and funded. This is going to require your audience to take a risk, so you want them ready to cross the threshold and sign off on your idea.

One more quick tip on slide decks: bullets get boring. Mix it up by interspersing visual queues, graphics, symbols, and other interactive elements to break up the flow. If your audience gets bored or distracted, you lose them.

6.5 MAKE A MEANINGFUL IMPRESSION

"THE MOST OBVIOUS CLUE WAS SARTORIAL: CLEANTECH EXECUTIVES WERE RUNNING AROUND WEARING SUITS AND TIES. THIS WAS A HUGE RED FLAG, BECAUSE REAL TECHNOLOGISTS WEAR T-SHIRTS AND JEANS.

THERE'S NOTHING WRONG WITH A CEO WHO CAN SELL, BUT IF HE ACTUALLY LOOKS LIKE A SALESMAN, HE'S PROBABLY BAD AT SALES AND WORSE AT TECH."

PETER THIEL [67]

"Never judge a book by its cover" is a mantra absolutely no one follows. We all make assumptions based on our first impressions, whether we're cognizant of it or not. When you meet with investors — either casually at an industry event or at an actual sit-down — they will judge your product based on your person.

Research suggests, and empirical evidence shows, that the effects of appearance on social outcomes are pervasive. Like all of us, investors look at us and make judgments in milliseconds about our status, competence, likeability, trustworthiness, and aggressiveness.

---

67 Peter Thiel and Blake Masters, *Zero to One* (Penguin Random House, 2014).

They make these assumptions rapidly, and effortlessly, in less than the blink of an eye — literally — and research has shown that such snap judgments are often accurate.[68] As they get to know you, people often gain confidence about their judgments and, occasionally, change their mind. That's a risk you can't take with investors.

ONE OF THE SIMPLEST THINGS YOU CAN DO TO IMPROVE YOUR CHANCES OF SUCCESS AND SECURE THE APPROVAL OF INVESTORS IS TO MAKE AN UNFORGETTABLE FIRST IMPRESSION.

This might sound obvious, but you'd be surprised how many people will spend hours and hours perfecting a presentation deck, yet virtually zero time attending to their personal appearance before they go in front of an investor. If you're presenting to a well-respected angel firm, don't show up in an untucked, ill-fitted shirt with stains on it.

On the other hand, it's important to match your appearance to the culture you're presenting to. If the culture of the investor that you're meeting with is, in fact, untucked, then don't show up in a starched shirt and a conservative suit.

THERE'S A BALANCE TO BE STRUCK. STRIKE IT.

6.6 CONSIDER YOUR ODDS: TO VC OR NOT TO VC

"FINISH EACH DAY AND BE DONE WITH IT. YOU HAVE DONE WHAT YOU COULD. SOME BLUNDERS AND ABSURDITIES NO DOUBT CREPT IN; FORGET THEM AS SOON AS YOU CAN. TOMORROW IS A NEW DAY. YOU SHALL BEGIN IT SERENELY AND WITH TOO HIGH A SPIRIT TO BE ENCUMBERED WITH YOUR OLD NONSENSE."

RALPH WALDO EMERSON

---

68 Nicholas Rule, "Snap Judgment Science: Intuitive Decisions About Other People," *Association for Psychological Science Observer*, May/June 2014, http://www.psychologicalscience.org/observer/snap-judgment-science#.WV-0uNPytUN

Given the rampant success of tech startups — not to mention their cultural stereotyping — you might picture VCs flush with cash investing indiscriminately in endless ideas. The juggernaut tech companies such as Google, Amazon, Apple, Slack, and Dropbox all started as good ideas with solid funding. But don't be fooled; the competitive environment for technology products is extreme. It's not as easy to get funding as TV shows like *Silicon Valley* would have you believe. Although many VCs are indeed oozing cash, they are getting more and more rigorous in their vetting.

Even if you have the most exciting, innovative product idea in the world — something no one has ever done before — you're going to have a hard time landing VC funding. When Tim Westergren ran out of money from his initial seed round (which he received when all seed funding required was a business plan), he was able to keep fifty employees working for two years while he pitched what would eventually become Pandora a total of 348 times.[69]

Westergren learned about failure early as an aspiring musician and composer, so he took each "no" as an opportunity to improve. With $500,000 in personal debt and $2M in back salaries, Westergren finally secured a $9M funding round that resulted in a pivot with his magical musical taxonomy and algorithmic matching engine into the launch of Pandora in the fall of 2005.[70]

69 "Founder of Pandora on Lessons from a Near Dot Com Bust to Million Dollar IPO," FirstRound.com, accessed February 16, 2018
http://firstround.com/review/Founder-of-Pandora-on-lessons-from-near-dot-com-bust-to-billion-dollar-IPO/
70 Matt Weinberger, "Pandora didn't pay 50 employees for two years," *Business Insider*, April 24, 2015,
http://www.businessinsider.com/pandora-didnt-pay-50-employees-for-two-years-says-tim-westergren-2015-4

This was a product created with a lot of blood, sweat, and tears — not the traditional sexy startup story. The timing was impeccable, though, because at the same time, Steve Jobs was about to revolutionize the smartphone. The introduction of the iPhone fueled rampant growth for Pandora.

Westergren's isn't an isolated case. There are many stories of entrepreneurs being rejected multiple times a day for months or even years before finally securing a seed round of funding. Brian Chesky, co-founder of Airbnb, once posted a *Medium* piece called "7 Rejections" with screenshots of all rejection emails he got from initial investor queries who didn't quite get the vision. Joke's on them!

In today's landscape, we recommend using VC rejections as a catalyst to strengthen your product. Unless you can impress and convince them with one of the following three things, the odds of rejection are high — even with an amazing product.

1. Existing users who love your product

2. Existing revenue that keeps the lights on

3. Impressive industry connections — you've either already launched a successful product or you know someone who has who will vouch for you

If you have none of these things, we're not saying you shouldn't approach investors. But have a realistic mindset. Your efforts will most likely fall under the category "good practice" until you do have one of the above assets. If you're hoping for utter enthusiasm, you're not going to get it. Our experience has shown that it's good to go in expecting a "no" and observe closely as you present to identify nuances or hesitancy from investors. Hesitancy is actually a good sign that you are onto something.

MOST TRULY REVOLUTIONARY IDEAS ARE NOT IMMEDIATELY ACCEPTED, BECAUSE ANYTHING TRULY NEW REQUIRES A SHIFT IN THE INVESTORS' MENTAL MODEL.

The reality is that most VCs are looking for the next 10x revenue generator, a product that will be a driver of revenue for them. And most of these firms do not have balanced portfolios. In many cases, the most profitable company within a VC's portfolio is more profitable than the rest of the portfolio combined. Understanding this concept is foundational to how you position and pitch your company for funding. (And if a VC is actively pursuing your company, consider yourself lucky. This is the ideal position to be in for negotiation.)

AN IDEA BEFORE ITS TIME

Many years ago we spotted an exciting new opportunity for a very high-profile client. Based on an initiative this client had launched, we started hearing from bloggers around the country who wanted to be involved. This gave us an idea. What if we created an interactive widget that we could place on the blogger sites that would give their users a similar experience they'd get on our client's main site? We envisioned videos that would play inline, as well as a form a viewer could submit right from the widget to sign up. Then we built such a thing, which back then, was unheard of.

Almost immediately, the data proved that our hunch was right. These blogger widgets were getting more engagement than our client's main site. This was early days influencer marketing, although it didn't have a name back then, and it worked.

Yet, even with the data to show for it, we had a hard time getting our client and colleagues to expand on the idea, or even pay attention. We'd start to explain, and they'd check out. Even with analytics, decks, and analogies, it was nearly impossible to find anyone who could catch the vision. Unilaterally, the idea was rejected by everyone around us: designers, developers, products managers, account directors, and analysts. Perhaps we did a poor job selling it, or maybe it was an idea before its time, too novel in the moment.

We believed in our idea, even though at times it felt like a fool's errand, so we kept pursuing it in our spare time. This widget was quietly driving more engagement than hundreds of thousands in ad spend. We had to ask ourselves, "Are we crazy? Why is it we are the only ones that see the value here?"

Months and months later, through sheer force of will and relentless perseverance, we received funding to run a test for a holiday campaign. This was the jumpstart the project needed, and the product is now thriving. Today, such influencer relationships are now a commonplace type of digital marketing.

All good ideas at first sound crazy and sometimes seem impossible to pull off. That's why big companies, married to their processes and safety zones, are not often the innovators.

ARE YOU THE EXCEPTION?

"ARTIFICIAL INTELLIGENCE HAS BECOME THE MOST HYPED SECTOR OF TECHNOLOGY. WITH NATIONAL PRESS REPORTING ON ITS DRAMATIC POTENTIAL, LARGE CORPORATIONS AND INVESTORS ARE DESPERATELY TRYING TO GET INTO THE SPACE." — RIVA-MELISSA TEZ ON MEDIUM[71]

If you're reading this chapter thinking "my idea is the exception to this rule," you might be right — if your product is an artificial intelligence (AI) product. AI is the hot ticket for startups these days, which is why the international NIPS conference (stands for Neural Information Processing Systems) has grown exponentially in recent years. Held in Barcelona in 2016, NIPS drew AI researchers, developers, and thought leaders from every major and minor player worldwide, including teams from Google's DeepMind and Brain, Facebook AI Research, MIT, and Stanford — plus hundreds of startups.

---

71 Riva-Melissa Tez, "Rocket AI: 2016's Most Notorious AI Launch and the Problem with AI Hype," *Medium: The Mission*, December 15, 2016, https://medium.com/the-mission/rocket-ai-2016s-most-notorious-ai-launch-and-the-problem-with-ai-hype-d7908013f8c9#.hyqpgq8sp

In 2015, 33 AI companies were acquired, adding to the hype. If your company likes to throw around terms like machine learning, reinforcement learning, or deep learning, you might very well be riding this wave.

Interesting sidebar to this sidebar: In 2017, a Bloomberg report revealed that some of Google's best AI talent recently left the company after receiving such huge bonuses — presumably as an incentive to stay with Google — that they were able to retire.[72]

TIMING IS EVERYTHING

Let's take a step back for a moment. VCs want you to believe you need their funding, but it's a very personal decision to determine if and most importantly when to try for it. If you ask a VC, they'll give you multiple case studies that prove the necessity of having funding, and they'll highlight their successful stories of startups that struck gold. But they won't tell you about the overwhelming majority of companies that not only did not strike gold, but ultimately went broke and lost everything. Another secret they like to keep off the radar is that many of these failures include VCs themselves.

You don't absolutely need venture funding in order to launch a product.

OUR BELIEF IS THAT IT'S GENERALLY BEST TO ESTABLISH YOUR PRODUCT IN A NICHE MARKET FIRST, IF YOU CAN.

This can take time, but it allows you to solidify your core team and product. While VC funding can be critical to expanding and capitalizing on market timing or a specific opportunity, we still believe there is opportunity for self-funded startups.

TALK ABOUT FAILING SPECTACULARLY

Throughout this book, we've championed the mindset of embracing failure, and we stand by that belief. But there's one area in which you might not want to fail too spectacularly, and that's with your VC funding.

Over the past few decades, there have been some colossal failures in the tech world, like CueCat, a barcode scanner that was supposed to revolutionize e-commerce.

---

72 Ben Kew, "Report: Google Self-Driving Car Designers Leave Company After Receiving Huge Bonuses," *Breitbart News*, February 14, 2017, http://www.breitbart.com/tech/2017/02/14/report-google-self-driving-car-designers-leave-company-after-receiving-huge-bonuses/

The parent company, Digital Convergence Corp., raised $185 million and flopped entirely in 2001 because no one wanted to use the devices. Mode Media raised $229 million, but shut down due to financial mismanagement. Quirky was financed to the tune of $185.3 million by elite investors such as Andreessen Horowitz, but this startup floundered too.[73] The list goes on.

In fact, Forbes reported that about 55 percent of one VC firm's capital went to losing deals over both a 10- and a 30-year period — meaning that their track record has not improved. And Harvard Business School lecturer Shikhar Ghosh analyzed a 2012 study of 2,000 companies that received at least $1 million in venture funding to unearth the disturbing fact that 95 percent of startups don't yield their projected returns. Some predict that loss ratios are going to get worse, as VCs continue to champion the "fail fast" mentality while still throwing outrageous sums of money at problems.[74]

So yes, we urge you to fail fast. But consider getting some of that failure out of the way before it involves millions.

On the other hand, as a new product developer, you might have one small challenge with being fund-free: coming up with the money to build out your team and office space. In these days when everyone seems to work at home, this concept can be easily overlooked.

BUT TO RECRUIT WELL, YOU NEED TO SET THE STAGE, AND YOUR PHYSICAL SPACE IS AN EXTENSION OF YOUR VISION, MISSION, AND PRODUCT.

The cost for skilled engineers and other talented folks continues to grow and outpace other wages. Having the right space can be essential to helping talented people differentiate between two good opportunities. The key thing to remember is it may be far cheaper to invest in the "right" office space than try and compete on over-inflated salaries. This becomes especially true as you grow.

So if you decide that it's time to pursue funding, good luck. And when you get it, we have one simple recommendation to offer: please don't blow it. Don't go crazy and spend it all on arcade games and Eames furniture. It's so common for VC-backed companies to go on spending binges to improve their company culture in order to reward the team and attract bigger talent. But off-the-wall spending usually results in bad habits, unsustainable expenses, and real failure. Even with fat funding, you'll eventually go broke and run your business into the ground if you don't use it intelligently.

---

73 "121 of the Biggest, Costliest Startup Failures of All Time," *CB Insights*, November 10, 2017, https://www.cbinsights.com/blog/biggest-startup-failures/
74 Maheed Zaidi, "Billions in VC funding deals could be lost due to a high tolerance for failure," *Tech Portfolio*, Accessed July 7, 2017, http://techportfolio.net/2016/08/vc-funding-and-the-culture-that-rewards-failure/

7.0 THE RIGHT AND WRONG TIME TO QUIT YOUR DAY JOB

We know, you're excited. But before you make any rash decisions about quitting your day job, stop and think about the consequences. The key is knowing when to make the leap.

If you're really serious about quitting, it's an incredibly personal decision that we cannot make for you. We can, however, pose some questions for you to ask yourself:

ARE THERE OPPORTUNITIES IN YOUR CURRENT SITUATION YOU'VE OVERLOOKED?

You may already be in a role at your current job where you can apply the ideas you read about in this book. Being an "intrapreneur" — someone who innovates from inside an existing organization, developing new, blue-sky products with an existing team — is far easier than striking out on your own. You may already have the resources, human and otherwise, and the support you need to be bold without having to raise the white flag and leave altogether.

This can be a first line of defense, allowing you to stay within the comfort zone of your salary and routine while exploring your entrepreneurial side.

IF YOU WERE TO QUIT YOUR JOB TODAY, COULD YOU STILL PAY YOUR BILLS?

Along with self-growth, you of course have to think about, well, money. Obviously, if you couldn't pay the bills without your day job, you probably shouldn't quit just yet. You might have to organize your life a little differently. Regardless of your zest for entrepreneurship, you have to be real about your debts and your dependents. When you get to a point where you can support yourself and those who rely on you perfectly well for a while without that day job, then you can think about quitting. And don't underestimate the time it might take to get a product off the ground and profitable.

Remember, though, a day job is more than just a paycheck. It's a culture, a way of life with lots of benefits besides cash money. Are you ready to give up that culture and those benefits? For instance, leaving a full-time position usually means leaving health insurance and paid time-off. You need to take all of these things into account before making the leap.

IS YOUR PROJECT STAGNATING BECAUSE IT NEEDS WAY MORE LOVE AND ATTENTION?

You've been building a product on the side, and you're starting to run out of free time. Your project is building momentum, and you know it's ready for full-time — and your day job is suffering from your split attention.

If quitting your day job enables you to increase the number of clients and users for your product, it might indeed be time to jump. Stagnation can quickly kill the energy around a product. Since you are the vision holder, you have to keep the passion and energy alive with your product. Nobody but you can do this; you can't just hire for it.

IS YOUR WINDOW OF OPPORTUNITY CLOSING?

Luck is when preparation meets opportunity, as they say. Some products have built-in windows of opportunity and a limited timeframe for success, because of either market factors or competition. If you need to give your full attention to your product in order to bring it to market during a crucial window, that's something to consider when weighing a decision to quit your day job. Day jobs can be a distraction and can ultimately interfere with your success.

However, in many organizations, valued employees have access to less drastic options than abrupt departure. Extra vacation days, special project approvals, leaves of absence — these can be creative ways for you to keep your job without risking a leap into a scary abyss of unemployment.

PERSONAL GOALS

Your own personal fulfillment and life goals are factors you shouldn't always ignore in favor of practicality. All too often, it's easier for us to maintain the status quo than to follow our dreams. The barriers to getting out of our current situation (e.g., job) might seem too enormous to surmount, but as Anaïs Nin famously said, "The day came when the risk to remain tight in a bud was more painful than the risk it took to blossom."

There is a risk to staying in the status quo. It's always easier to look back and connect the dots, but if you find that you aren't waking up and enjoying what you do, maybe you should go for it. If you've tried being intrapreneurial, still feel boxed in, can't progress in your current environment, and are crippled by a micromanager boss, perhaps it's time.

LIFE SCORECARD

TIMES WHEN I THOUGHT...

"I'M NOT REALLY HAPPY HERE, BUT MAYBE THIS IS THE BEST I CAN EXPECT AND I'LL REGRET GIVING UP."

... IT TURNED OUT I...

SHOULD HAVE STAYED	SHOULD HAVE LEFT SOONER
II	HHH HHH III

Just be sure you've weighed all your options, considered every factor, and more than anything, that you check in with your gut. Don't mistake a bad day or a bad week or a sucky project for the wrong job.

If you've explored all the options within your current environment, this level of self-awareness will help you to decide when it's time to make the leap, not just acting rashly because you're excited, but making informed, intuitive decisions based on your real life.

THERE IS NO EXACT FORMULA WE CAN GIVE YOU FOR WHEN TO QUIT YOUR DAY JOB, BUT HOPEFULLY, HAVING THE ABOVE DISCUSSION WITH YOURSELF WILL HELP GIVE YOU A LITTLE MORE CLARITY.

The Value of a life Scorecard

- Justin Jones -

It's often true that when an employee is unhappy in a job, so is a manager. Once, a friend of mine hired his brother for a position that was simply not the right fit. This "favor" ended up being a curse. The brother struggled in the job, and their relationship began to sour. It wasn't comfortable for either of them, but my friend did not want to fire his own brother, and the brother felt too guilty to quit. Instead, they both became more and more frustrated.

Had the brother been keeping a scorecard, he would have recognized that it was time to leave and pursue a career path better suited to his passions. Eventually, things came to a head, and my friend was forced to fire his brother.

While this situation seemed strained at the time, in the end, it was the best thing that could have happened to either of them. The brother went on to become completely successful in a different line of work.

I know of at least a half-dozen relationships like this where the employee is unhappy in a role, and the manager unhappy with performance, but they both stay put because of a misguided sense of loyalty, kindness, or fear. If a poor relationship with your boss and a bad fit at work are affecting your personal life, you know it's time for a change.

8.0 COMPETING WITH BUREAUCRATS

"IN THE MOST DYSFUNCTIONAL ORGANIZATIONS, SIGNALING THAT WORK IS BEING DONE BECOMES A BETTER STRATEGY FOR CAREER ADVANCEMENT THAN ACTUALLY DOING WORK (IF THIS DESCRIBES YOUR COMPANY, YOU SHOULD QUIT NOW)."
PETER THIEL [75]

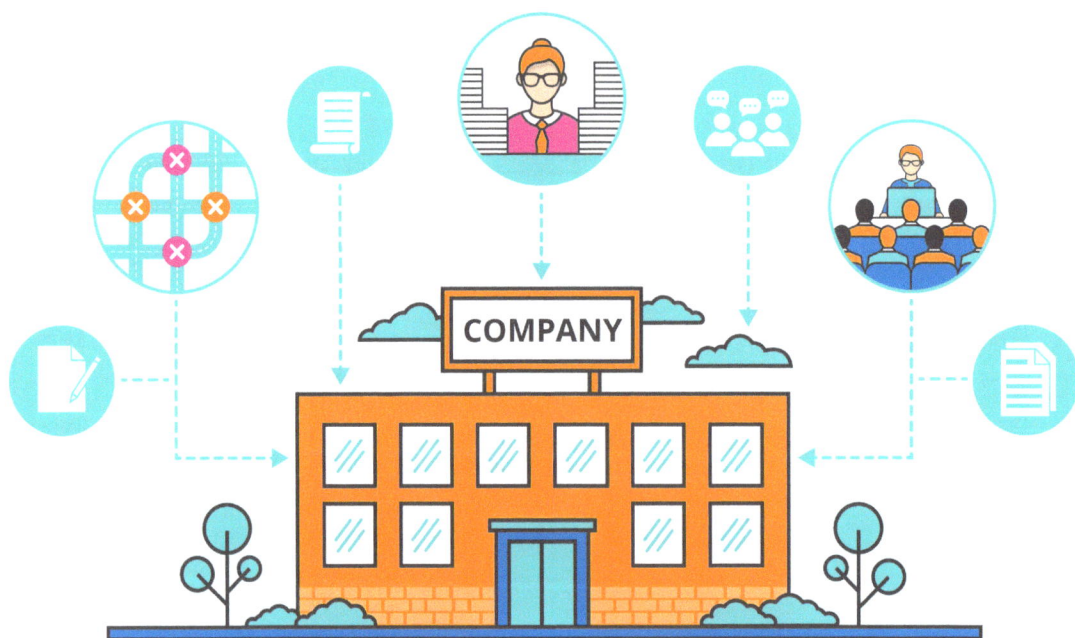

If you still have a day job (or ever have), you are probably familiar with the bureaucrat archetype. The larger the organization, the more likely you are to be surrounded with bureaucrats and dysfunctional teams, departments, and divisions. Or if you are running a business, you may have clients or partners that are bureaucrats.

Where bureaucrats have infiltrated an organization, an insidious form of egalitarianism rules the day. An informal doctrine exists where all people deserve equal rights, opportunities, and pay based sheerly on tenure and title, as opposed to their actual impact on the organization or its products.

---

75 Peter Thiel, *Zero to One: Notes on Startups, or How to Build the Future* (2014, Crown Business).

Unless they are in traditional commission-based sales roles, top performers are paid within the same range as co-workers who produce significantly less impressive results.

THIS STATUS QUO BREEDS STAGNATION AND COMPLACENCY,
WHICH IS COMPOUNDED BY THE HERD EFFECT.

In such organizations, the flow of data goes up to the top of the org chart to executive committees. The competing factions on these politically motivated committees then vacillate and negotiate for periods far longer than necessary. Most of the time, the decision is left to the HiPPO (highest paid person's opinion — an acronym coined by Amazon in the early 2000s and now widely adopted as a practice to avoid in tech companies). This is a recipe for slowness and failure.

It's also the antithesis of our mindset, which is all about speed, agility, testing, and iteration. We believe ideas should be held up to their merit. It's so cheap and easy to test product ideas that it's idiotic not to. It's far cheaper, in fact, to run a test at ground level, where stuff gets built, and glean data from that test, than it is to hire executives to sit in a boardroom and pontificate over that idea.

BASED ON OUR OWN EXPERIENCE, IF WE HAVE TO SIT IN ONE MORE BOARDROOM AND LISTEN TO A HARVARD GRAD WITH IMPRESSIVE EDUCATION BUT LITTLE REAL-LIFE EXPERIENCE SPOUT OFF HIS UNINFORMED OPINIONS ABOUT PRODUCTS...

Anyway, when a decision is finally reached in a bureaucratic organization, it has to make its slow way back down the chain of command. This structure ensures that those in charge are always in control; it also slows everything to a crawl and tends to water down ideas.

Bureaucrats are data filters and decimators of information, overly concerned with procedural correctness and organizational processes at the expense of the needs of people or products. They prioritize process, and their place in the organizational hierarchy, over results.

Bureaucrats like to move slowly and carefully in a hopeless attempt to get committees of people on the same page, where self-interest trumps organizational goals and product vision.

BUREAUCRATS ARE RISK-AVERSE. IT'S INEVITABLE YOU ARE GOING TO WORK WITH THEM, AND YOU MAY EVEN HAVE TO COMPETE WITH THEM.

THE CIA'S TIMELESS GUIDE TO SABOTAGE FROM WITHIN

In 1944, in the heat of World War II, the government organization that would soon become the CIA (then called the Office of Strategic Services) released a document called "The Simple Sabotage Field Manual." The purpose of this document and its distribution was to empower dissenters in enemy countries to undermine their own governments from within. Funnily enough, this guide could easily be a "how to be a bureaucrat" manual. It's uncanny how these bullet points align with both enemy dissenters and well-meaning bureaucrats.

1. Managers and Supervisors: To lower morale and production, be pleasant to inefficient workers; give them undeserved promotions. Discriminate against efficient workers; complain unjustly about their work.

2. Employees: Work slowly. Think of ways to increase the number of movements needed to do your job: use a light hammer instead of a heavy one; try to make a small wrench do instead of a big one.

3. Organizations and Conferences: When possible, refer all matters to committees, for "further study and consideration." Attempt to make the committees as large and bureaucratic as possible. Hold conferences when there is more critical work to be done.

4. Telephone: At office, hotel and local telephone switchboards, delay putting calls through, give out wrong numbers, cut people off "accidentally," or forget to disconnect them so that the line cannot be used again.

5. Transportation: Make train travel as inconvenient as possible for enemy personnel. Issue two tickets for the same seat on a train in order to set up an "interesting" argument.[76]

---

76 "Timeless Tips for 'Simple Sabotage'," Central Intelligence Agency, last updated April 30, 2013, https://www.cia.gov/news-information/featured-story-archive/2012-featured-story-archive/simple-sabotage.html

"HIPPOS ARE LEADERS WHO ARE SO SELF-ASSURED THAT THEY NEED NEITHER OTHER'S IDEAS NOR DATA TO AFFIRM THE CORRECTNESS OF THEIR INSTINCTUAL BELIEFS. RELYING ON THEIR EXPERIENCE AND SMARTS, THEY ARE QUICK TO SHOOT DOWN CONTRADICTORY POSITIONS AND DISMISSIVE OF UNDERLING'S INPUT."
CHRIS DEROSE AND NOEL TICHY, FORBES[77]

Bureaucrats come in a variety of flavors, but the two key stereotypes are:

1. The "Critic" who is managed by fear of failure and can only deny ideas.

Constantly covering his own ass, he's a roadblock to progress; and he always wants the "committee" to weigh in. He believes your ability to contribute is measured by your job title and position in the organization. He specializes in kicking the can down the road and thinks talking about work is more important than actually doing it. He is risk-averse, scared of making a mistake or being wrong.

2. The "Talker" who aspires to climb the organizational ladder.

Slightly less of a roadblock, but very eager to take credit for your hard work. He doesn't know what he doesn't know, and never admits failure. He wants you to work behind the scenes and feed him ideas that increase his perceived value to the organization (and we say "him" because this is the blueprint of the alpha-male ladder climber). He carefully disseminates and hoards institutional information, and his power play is in the hoarding of strategic information and relationships.

Because of the rigid focus on process, at times, the bureaucratic type can feel like the enemy — especially if their focus on process comes at the expense of your product. We would challenge your pessimism around dealing with bureaucrats. Not everyone has the entrepreneurial spirit, but everyone adds value to your startup project — even bureaucrats. It's time to reframe the situation.

Any marketer will tell you: "Know your audience." Knowing your audience helps you speak to them in a way that resonates. So if you're speaking to bureaucrats, speak to them in their language. We know you care more about your product then the process, but you still need a process to deliver your product. When talking to bureaucrats, talk about the process and the value they bring within it.

---

77 Chris DeRose and Noel Tichy, "What Happens When a 'HiPPO' Runs Your Company?," *Forbes*, April 15, 2013, https://www.forbes.com/sites/derosetichy/2013/04/15/what-happens-when-a-hippo-runs-your-company/#775ab88b40cf

Working together with bureaucrats is the first step. Then, it's time to leverage them. This requires setting aside your own ego, having a long view, and playing the game. The bureaucrats have already played their hand. They expect entrepreneurs and intrapreneurs to act in a certain way — to disregard the rules and processes. Don't play into these expectations. Instead, use the rules and processes to your advantage, and give bureaucrats a role. Be a wolf in sheep's clothing: disguise yourself as one of them by placing them within the process. Understand their ambitions, anticipate how they will behave and what they will say in a given situation, and then leverage this knowledge to get your project approved and off the ground.

"THE QUESTION ISN'T WHO IS GOING TO LET ME; IT'S WHO IS GOING TO STOP ME."
AYN RAND, WIDELY PARAPHRASED

Remember, your objective is to get your product built. How you get that done is not important. Don't let your ego get in the way. Compromise is the key to getting a successful product developed.

THE DOWNFALL OF BIG ORGS

In larger organizations you run into many challenges, but there are two key ones to be aware of:

1. There are a lot of grandiose plans and visions for the future of the product, company, and deliverable. This is unavoidable, because in order to profit, large organizations need a big market and lots of users for their products. But the downside of this is that they lose site of the idea of small incremental wins.

2. The bigger the organization, the further in advance it tends to plan and budget. Often, big companies do their planning and budgeting a year or even years in advance. This practice is fundamentally at odds with a startup mentality. Instead of being nimble, it creates inflexibility and resistance to change. Innovative ideas don't sync with the plan, and trying out new things is discouraged. This is an innovation killer.

9.0 THE MIRACLES OF FAILURE

Throughout this book, we've talked about how important it is to be willing to fail, and we want to end by reminding you of this crucial mindset. For inventors, creators, and innovators, it's perhaps the most important attribute one can have.

Creativity expert Sir Ken Robinson, in his brilliant TED Talk "Do Schools Kill Creativity?," talks about how we're all conditioned to fear failure in our educations. The public school system breeds a willingness to fail right out of us, and for most of us, our professional lives uphold this edict: DON'T FAIL.

> "FAILURE SHOULD BE OUR TEACHER, NOT OUR UNDERTAKER.
> FAILURE IS DELAY, NOT DEFEAT. IT IS A TEMPORARY DETOUR, NOT A DEAD END.
> FAILURE IS SOMETHING WE CAN AVOID ONLY BY SAYING NOTHING,
> DOING NOTHING, AND BEING NOTHING."
>
> **DENIS WAITLEY**

Failure, in our opinion, is a word that has an unjust stigma. A failure is the result of an incorrect decision, and is often a path to success. We love the word *failure*, because the more we fail — and the quicker we fail — the brighter that path to success is lit.

Think of yourself as a scientist. When scientists propose a hypothesis, they investigate it with as objective of a mindset as possible. You, too, should be objective. If your product does not work out, you can then say "We investigated our hypothesis, but discovered that the data did not support it, so we are currently looking for a new one." Disproving a hypothesis is part of the scientific journey. And failure is a part of the product journey.

"AN EXPERIMENT GONE WRONG DOESN'T HAVE TO MEAN SOMEONE GOOFED. IN A CULTURE OF GROWTH, IT SHOULD MEAN THAT YOU TRIED SOMETHING NEW, MEASURED THE RESULTS, AND LEARNED THAT THE CHANGE DIDN'T HELP THE BOTTOM LINE." — CASEY CAREY, DIRECTOR OF GOOGLE ANALYTICS MARKETING[78]

In a Think with Google 2017 newsletter, Google's Director of Analytics Marketing suggested that companies would be well served to publish a "quarterly failure report." The idea is to glorify what's learned from mistakes, particularly during testing — anything from a button that doesn't work to an abysmal checkout flow.

At Google, Casey Carey believes in reinforcing a culture of failing, and failing fast. After all, the faster you fail, the faster you learn from failure. And if you're not failing, it means you're playing it too safe with your testing.

When talking specifically about building products, failure simply means that the product needs to go through another iteration. Therefore, you want to surround yourself with people who are willing to learn from failure, and purge your team of those who are afraid of it. Fear of failure is like a boat anchor that will bring your product to a grinding halt before it reaches its destination.

Jeff Bezos, founder of Amazon, has always been a fan of failure. He has said about Amazon:

"I BELIEVE WE ARE THE BEST PLACE IN THE WORLD TO FAIL (WE HAVE PLENTY OF PRACTICE!), AND FAILURE AND INVENTION ARE INSEPARABLE TWINS."[79]

But he has also pointed out that a willingness to fail is contrary to human nature. While people want to be inventive, they are embarrassed if they fail. Yet true inventiveness requires that one get out of their comfort zone and be able to fail gracefully, and often.

---

78 Casey Carey, "Why Every Marketer Needs a Quarterly Failure Report," *Think with Google*, January 2017, https://www.thinkwithgoogle.com/articles/benefits-of-failure-marketing-analytics.html
79 Monica Nickelsburg, "In annual letter, Jeff Bezos says Amazon is an 'invention machine' defined by failure," *GeekWire*, April 5, 2016, http://www.geekwire.com/2016/annual-letter-jeff-bezos-says-amazon-defined-failure-good-thing

When pressed about how his company has experienced failure, Bezos cites the example of the Amazon smartphone, dubbed the Fire Phone when it debuted back in 2014. The Fire Phone was an enormous flop, and Amazon wrote off $170 million because of it.[80] But Bezos loves to use the Fire Phone as an example of the company's spirit of innovation:

IT IS OUR JOB, IF WE WANT TO BE INNOVATIVE AND PIONEERING, TO MAKE MISTAKES AND AS THE COMPANY HAS GOTTEN BIG — WE HAVE $100 BILLION-PLUS IN ANNUAL SALES, 250,000-PLUS PEOPLE — THE SIZE OF YOUR MISTAKES NEEDS TO GROW ALONG WITH THAT.[81]

In Steve Jobs's famous Stanford University commencement speech of 2005, one of the most shared speeches of our era, Jobs spoke about how his early decisions didn't seem to make a lot of sense at the time, but looking back, the dots all connected. He operated largely from intuition, and the success of his products and the company he founded with partner Steve Wozniak, Apple, owes a lot to that intuition — and to his willingness to try crazy new ideas that might fail.

JOBS'S FINAL WORDS TO THE STANFORD GRADUATING CLASS STICK WITH US TO THIS DAY: "STAY HUNGRY, STAY FOOLISH."

It's no accident that Jobs and Bezos have shared a commitment to failure. This is the hallmark of a true innovator. The miracle of failure, in plain language, is this: First, failure lights up the path to success. Second, those who embrace failure are individuals you want on your team. By embracing failures, recording them, and studying their meaning, your path to success will be illuminated.

---

80 Taylor Soper, "Ouch: Amazon takes $170M write-down on Fire Phone," *GeekWire*, October 23, 2014, http://www.geekwire.com/2014/amazon-takes-170m-loss-fire-phone/
81 Monika Nickelsburg, "Amazon's Jeff Bezos on the Fire Phone: 'We're working on much bigger failures right now,"*GeekWire*, May 19, 2016, http://www.geekwire.com/2016/amazons-jeff-bezos-fire-phone-working-much-bigger-failures-right-now/

10.0 THE PLAYBOOK TO GET 'ER DONE

STOP TALKING.
START ITERATING.

ALWAYS QUESTION
THE STATUS QUO.

BE RELENTLESS
WITH YOUR
VISION.
NEVER SETTLE.

LISTEN BEYOND
THE DATA TO
UNDERSTAND
WHY.

ALWAYS KEEP
YOUR USER AT
THE CENTER.

FOCUS ON SPEED
OF ITERATION,
NOT PERFECTION.

BECOME YOUR
OWN FIXER-UPPER
IN EVERY WAY.

CREATE A DO-NOT
LIST (TIME IS MONEY).

INSPIRE OPTIMISM
WITH YOUR EMPATHY.

11.0 22 TIPS TO MARKET YOUR PRODUCT

ONE LAST THING — MARKETING.

BUILDING AN AMAZING PRODUCT IS ONLY HALF OF THE EQUATION. THE OTHER HALF IS REALLY UNDERSTANDING HOW TO MARKET IT.

How will your potential users find your product? How are you going to tell them about it? Good marketing is essential to your ability to have a successful product.

You might not be thinking about marketing yet, but it's important that you do, because it has to be part of your plan and budget. If you spend all your money developing a product and don't have any left over to promote it, you'll kill it. There are exceptions to this rule, but, they are definitely exceptions.

We could write an entire book on how to market a product (and, in fact, we did), but for now, we want to leave you with a few pointers to get you thinking. Some of these things can be done very early on, and some will have to wait until the product is ready to sell. But by reading about them now, you can work them into your plan and vision.

1. Kick off your marketing with some guerilla research on whether the name of your product is memorable, easily pronounceable, and communicates what you think it does. Make sure it's not a curse word in Chinese.

2. Pricing is always a guesstimate, but if you have built a superior product, then you should sell it for a premium price.

3. Always leverage friends, colleagues, college classmates, current customers, and everyone else you can reach out to, to the fullest possible extent. If you already have a relationship with someone, it's much easier to sell them something — and then they share with their connections, and you have a marketing machine in motion.

4. Remember the 80/20 rule: 20 percent of your customers will use and advocate for your product 80 percent of the time. Figure out early on who these users are, profile them (so you can look for more users just like them), leverage them, and thank them.

5. Think integrated. Modern marketing and user behavior starts with the phone as the new hub (it used to be TV). In fact, according to comScore's 2016 Mobile App Report, "Mobile has grown so fast that it's now the leading digital platform, with total activity on smartphones and tablets accounting for two-thirds of digital media time spent." Start identifying mobile touchpoints to connect and engage with users.

6. Humor is a potent selling tool, but it isn't easy, and it must be a part of a greater ad strategy. A lot of startups try to mimic the hilarity of Super Bowl ads but lack the budget and brand to pull it off. We don't recommend focusing on the humor angle. Humorous ads can be handy for building brand recognition and goodwill, but that doesn't always translate to sales.

7. Segment, segment, segment. Instead of trying to compete head to head with major brands in the digital space, spend some time up front finding niche segments you can target with your media plan.

8. Ads aren't created equal, the difference between some ads in terms of sells can be 100 to 1. Kill the losers — the ads and marketing campaigns that don't produce immediate results. Focus your time, money, brain power, and energy on what's working.

9. You know your product best. Instead of relying on outside firms, whenever possible, use the talented individuals you already have on your team to help communicate your vision.

10. A key to a successful ad is to promise the consumer a benefit: easier to use, less hassle, more efficient, saves time, costs less, gives a better experience, etc.

11. It's not about getting people to buy it, but to use it — and to use it more often than any other product in that genre.

12. Most effective ads span countries and continents. The same is true with products.

13. When it comes to marketing, avoid creating by committee. The result is usually complicated campaigns with dozens of objectives — all attempting to reconcile disjointed views of the committee. Simple and clear objectives are always better and more achievable, and those occur when one person is in charge.

14. When you find a campaign that works, don't try and reinvent it prematurely. Take a measured approach to incremental improvement and engagement. Build on the brand awareness.

15. For now, at least, the post office still exists. Take advantage of this! For some products, mail is still an incredibly efficient way to get sells.

16. Evaluate your current user base and use it to target prospects, but don't overthink it. Boil your detailed analytics down to two key elements:
    1. Recency — the more recently a person has used or purchased a similar product, the more receptive she will be to your marketing messaging.
    2. Frequency — how often a person uses or purchases similar products indicates the level of desire to purchase.

17. Unless you have a celebrity or an influencer endorsing your product, you need to view all social properties like Facebook and YouTube as ad networks. Organic social reach is a myth for most businesses. Invest in ads.

18. Emulate blogs and publications with large followings. Most people new to marketing a product are obsessed with self-promoting and doing what they think is best (not what is actually best to increase users of the product). Watch out for the ad guy that's trying to push his pet campaign.

19. Offer free trials and samples, but...

20. For the most part, promotions are for losers. Put your money into telling your story and creating brand/product awareness. Promotions are like heroin — it's easy to become addicted to the instant rush, but it's a nearsighted gain.

21. Use the power of the "thank you." Thank your customers, users, employees, beta testers, anyone you come in contact with — especially anyone who stood by you through the process of creating a new product.

22. Move fast. Review progress every two to three weeks. Do not fall into the trap of trying to measure success over long periods. You should know in two weeks or less whether a placement will be successful.

THIS IS JUST A SAMPLING OF WHAT TO EXPECT IN OUR NEXT BOOK,
GOT USERS? HOW TO PERSUADE PEOPLE TO NOT JUST BUY
BUT LOVE YOUR PRODUCT.
WE GO THROUGH A DEEP DIVE ON MEDIA PLANNING,
OPTIMIZATION AND PRODUCT MARKETING. AVAILABLE SOON!

GLOSSARY

AGILE A software development methodology that prioritizes individuals and interactions over processes and tools, based on the Manifesto for Agile Software Development (See Section 3.2)

CHILDREN'S ONLINE PRIVACY PROTECTION ACT (COPPA) A US law passed in 1998 to protect children under 13. Managed by the Federal Trade Commission (FTC), it restricts access by minors to any material defined as harmful on the internet, with the onus on website operators to discern whether they're collecting information from children under the age of 13. (See Section 2.3)

COMPONENT LIBRARY A group of design building blocks that might include basic UI elements like glyphs/icons, alerts, menus, drop-downs, labels, and form fields (sometimes called user interface (UI) kits, although we think they are different (See Section 1.6)

DATA-INFORMED Decisions made not exclusively based on data, but with data as one factor, and human intelligence and intuition as others (See Section 2.5)

DATA-DRIVEN Decisions made exclusively based on data, without taking into consideration human intelligence and intuition (See Section 2.5)

DESIGN SYSTEM A comprehensive design resource package that helps you create a fluid experience across platforms and devices, which gives you a foundational design to work from (See Section 1.6)

DEVELOPMENT METHODOLOGY (AKA SOFTWARE DEVELOPMENT MODEL) The model you choose to guide your team in building out your project (covered in this book: Waterfall, Agile, RAD) (See Section 3.0)

DIGITAL MILLENNIUM COPYRIGHT ACT (DMCA) Passed by Congress in 1998 to, in their words, "implement United States treaty obligations and move the nation's copyright law into the digital age." This act heightens the penalties for copyright infringement on the internet. (See Section 2.3)

FAKE-OUT AD An inexpensive, quick way to conduct market research by creating "fake" ads for your so-far undeveloped product, and using them to conduct A/B testing on things like your product name, your product description, your basic marketing verbiage, specific target markets, and calls to action (often on Facebook) (See Section 2.1)

FAMILY EDUCATIONAL RIGHTS AND PRIVACY ACT (FERPA) A US law that protects the privacy of student education records and gives parents protections around such things as report cards, transcripts, disciplinary records, contact information, and class schedules (See Section 2.3)

FEASIBILITY AUDIT A preliminary report you create to assess whether your product idea is viable technically, economically, organizationally, and legally (See Section 2.3)

FOCUS GROUP A fairly outdated method of conducting market research where you gather a group of folks together to discuss your potential product and unearth opinions, attitudes, and expectations (See Section 2.0)

GENERAL DATA PROTECTION REGULATION (GDPR) A European regulation that impacts any company that does business with any customer in Europe, the GDPR was enacted in 2018 to give citizens more control of their personal data and to simplify the regulatory environment for international businesses by unifying the regulations within the EU. (See Section 2.3)

GRID SYSTEM A system for placing design elements in a way that's orderly and pleasing, which helps translate the design into usable product code later (See Section 1.6)

HEALTH INSURANCE PORTABILITY AND ACCOUNTABILITY ACT (HIPAA) A US regulation that provides data privacy and security provisions for safeguarding patients' medical information (See Section 2.3)

MANIFESTO FOR AGILE SOFTWARE DEVELOPMENT (AGILE MANIFESTO) The basis of the Agile development methodology, this is a manifesto published in 2001 by a collective of 17 software developers in the spirit of creating a methodology that would prioritize individuals and interactions over processes (See Section 3.2)

MARKET RESEARCH A critical step in the product-design process that helps you identify your potential audience, narrow down ideas, identify your competition, and assess the state of the market (and where it has holes) (See Sections 2.0, 2.1, 2.2, 6.1)

MIND MAP An early-stage exercise in which you lay out your vision and write down everything you know about your theoretical product, typically on a whiteboard or using software like Lucidchart, for the purposes of sharing (See Section 1.3)

PERSONALLY IDENTIFIABLE INFORMATION (PII) Information that can be used on its own or combined with other information to identify, contact, or locate a person. This might include social security number, date and place of birth, mother's maiden name, and biometric records. This type of data is highly regulated. (See Section 2.3)

PLATFORM Both the device and the operating system your product will be built for, for example, any of the following: smartphones, tablets, the web, smart home devices, wearables like watches—and their accompanying operating systems (See Section 1.8)

PRODUCT GOALS What you hope your product will do and accomplish (besides landing you fame and riches and, hopefully, VC funding) (See Section 1.3)

PRODUCT VISION Your initial ideas for what your product will be, what problem it will solve for users, and how you will get it there (See Section 1.1)

PROOF OF CONCEPT Your early attempt(s) to put your product idea down on paper or make a very rudimentary prototype, with which you can sell your vision to potential users, employees, partners, and investors; can be crude sketches or a simple screen walk-through (See Section 1.2)

RAPID APPLICATION DEVELOPMENT (RAD) A type of development methodology that focuses on a streamlined, cyclical process of rapid iteration, with a focus on user experience (See Sections 3.3, 3.4, Chapter 4.0)

SOFTWARE DEVELOPMENT KIT (SDK) A set of software development tools to create an app or product to a specific software package, software framework, hardware platform, computer system, video game console, or other type of development platform (See Section 1.8)

SARBANES–OXLEY ACT (SOX) Passed in 2002, this act was a reaction to the rampant public accounting scandals of the early 2000s and mandated strict reforms to improve financial disclosures from corporations and prevent accounting fraud. Also called the Corporate Responsibility Act of 2002. (See Section 2.3)

SENSITIVE PERSONAL INFORMATION (SPI) See personally identifiable information

TOUCHPOINTS Specific tasks a user can engage with on and around a product, including all the various marketing channels within which a user might hear about a product before using it (See Sections 1.0, 1.5, 11.0)

USABILITY STUDY A test of your product's usability, usually conducted in a one-on-one setting and guided by a professional researcher, where you can observe users as they engage with your product or prototype (See Section 2.2)

USER INTERFACE (UI) KIT A downloadable template to help you quickly construct an app (See Section 1.6)

USER EXPERIENCE (UX) The experience your user has using your product and being exposed to your brand, and the most important element of building your product (See Chapters 1.0, 2.0)

USER GOALS What you hope your users will accomplish by using your product (See Sections 1.3, 1.8, 6.2)

USER JOURNEY MAP A graphical tool that helps identify gaps in your user experience and increases your ability to generate awareness about your product, enlighten customers to its benefits, and motivate them to take action (See Section 1.5)

WATERFALL A traditionally popular software development methodology which lays out a sequentially flowing and absolutely linear process of product development, from conception to initiation to analysis to design to construction to testing to implementation (See Section 3.1)

WORKFLOW DIAGRAM Also called "wireframe," this is an exercise to outline key activities, tasks, and workflows between your product's screens (See Section 1.4)

www.ingramcontent.com/pod-product-compliance
Lightning Source LLC
Chambersburg PA
CBHW051559190326
41458CB00029B/6478